Analog Circuit Design Techniques at 0.5V

ANALOG CIRCUITS AND SIGNAL PROCESSING SERIES

Consulting Editor: **Mohammed Ismail.** *Ohio State University*

(Continued after index)

Shouri Chatterjee
Kong Pang Pun
Nebojša Stanić
Yannis Tsividis
Peter Kinget

Analog Circuit Design
Techniques at 0.5V

 Springer

Shouri Chatterjee
IIT Delhi
New Delhi
India

Kong Pang Pun
Chinese University of Hong Kong
Hong Kong
China

Nebojša Stanić
Columbia University
New York, NY
USA

Yannis Tsividis
Columbia University
New York, NY
USA

Peter Kinget
Columbia University
New York, NY
USA

Series Editor:
Mohammed Ismail
Department of Electronic and Computer Engineering
Ohio State University
Columbus, OH 43210
USA

ISBN 978-1-4419-4354-5   e-ISBN 978-0-387-69954-7

Printed on acid-free paper.

9 8 7 6 5 4 3 2 1

springer.com

To world peace...

Preface

The field of analog circuit design has transformed itself over the past decades. Interest in integrating analog circuits side-by-side with digital circuits has led to the use of smaller and smaller devices, and lower and lower power supply voltages. The material presented in this book takes ultra-low voltage analog circuit design to a power supply voltage as low as 0.5 V.

The book discusses general design ideas and techniques for a supply voltage of 0.5 V. OTA design and biasing are covered extensively. We further present the design of analog systems using these techniques. They include the design of a filter with a PLL, a continuous-time $\Sigma\Delta$ modulator, a track-and-hold circuit, and receiver front-end circuits.

Much of the material presented in this monograph, originates in work done by Shouri Chatterjee for his Ph.D. at Columbia University. The continuous-time $\Sigma\Delta$ modulator was developed by Kong Pang Pun during his stay at Columbia as a visiting research scholar. The receiver front-end circuits were developed by Nebojša Stanić as part of work done for his Ph.D. at Columbia University.

We would like to acknowledge Tawfiq Musah, Ajay Balankutty, Edward Chiang and Junhua Shen (graduate and undergraduate students at Columbia University) for helping out with a variety of experiments and measurements. Many thanks to Professor Ken Shepard and Dr. Robert Melville of Columbia University for facilitating the use of various measurement instruments. Special thanks to Erwin Deumens of IMEC for the great support while taping out the filter and the continuous-time sigma-delta circuits. We also thank Analog Devices and Intel for supporting parts of this work. The track-and-hold circuit was fabricated with the generous support of Philips Semiconductors.

Finally, we would like to thank the staff of Springer for their helpful efforts during the writing of the book. The professionalism and enthusiasm of Katelyn Stanne (production editor) was especially appreciated.

IIT Delhi, New Delhi, India
Chinese University of Hong Kong, Hong Kong, China
Columbia University, New York, USA
Columbia University, New York, USA
Columbia University, New York, USA

Shouri Chatterjee
Kong Pang Pun
Nebojša Stanić
Yannis Tsividis
Peter Kinget
December 2006

Contents

1

Introduction

"Analog" is analogous to real life. In real life, most things are continuous with respect to time. Analog circuits provide the connection between the physical world and the digital computing signal-processing systems. As such the true power of digital signal and information processing can only be exploited if analog interfaces with corresponding performance are available.

Cost and size considerations push towards a co-integration of the analog interfaces and digital computing, signal processing on a single die. Cost and size considerations also push towards finer line widths. The International Technology Roadmap for Semiconductors [1] gives us a unique opportunity to look into the projected future of semiconductor technology and identify design challenges early. The line width of CMOS technologies is projected to keep scaling deeper into nanoscale dimensions for the next two decades. This will increase the functionality density, the intrinsic speed of the devices and thus the signal processing capability of the circuits. However, in order to maintain reliability, to avoid breakdown, to avoid thermal problems and to reduce power density, the maximum supply voltage has to be scaled down appropriately. Traditional analog circuit design techniques are well suited only for higher power supply voltages, where many devices can be stacked, and still there is enough voltage headroom for signal swings. In this book, circuit design techniques have been developed and demonstrated at ultra-low power supply voltages, typically at 0.5 V.

Figure 1.1(a) shows the projections for the supply voltage, based on [1]. The thick-oxide devices scale to tolerate a maximum supply voltage of 1.5 V. However, the use of thick-oxide devices requires an extra mask for fabrication, and are therefore costly. Supply voltage scaling of the thin-oxide devices is beneficial for digital circuits since it reduces the power consumption quadratically. Along with a reduction in the supply voltage, the transistor's threshold voltage, V_T, is reduced, but not as aggressively, in order to maintain good ON/OFF characteristics of the MOS transistors, and to reduce static leakage levels in digital logic circuits. A minimum standard V_T of about 0.2 to 0.3 V is foreseen. By about the year 2016, at the 22 nm technology node, a power supply voltage of 0.5 V is projected for low power digital circuits.

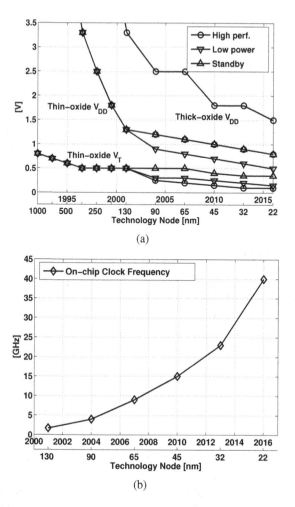

(a)

(b)

Fig. 1.1: (a) Supply voltage and threshold voltage scaling, (b) clock frequency scaling over time, as line widths decrease. The projections are based on [1].

The low power supply voltages and the relatively large device threshold voltages are an obstacle to high performance analog circuit design. Smaller supply voltages result in smaller available signal swing. To reduce circuit errors from thermal noise or offset voltages, often circuit power consumption has to be increased [2–5]. In addition, devices used in high speed linear circuits need to be biased in moderate or strong inversion with a minimum voltage over-drive, $(V_{GS} - V_T)$ of about 0.2 V, resulting in a $V_{DS,sat}$ requirement of about 0.15 V. Typical analog circuits require a supply voltage that is several $V_{DS,sat}$ plus the signal swing, or a V_T plus several $V_{DS,sat}$ plus the signal swing. At supply voltages below 1 V, the design of analog circuits becomes very challenging since the traditional circuit techniques do not have sufficient voltage headroom [4–9].

These challenges can be addressed with technology modifications or with circuit design solutions. A straightforward technology solution is to add thick oxide devices that are less aggressively scaled; these are slower but can operate, without breakdown, with larger supply voltages, as shown in Fig. 1.1(a). The other technology modification is to include low-V_T [10] or native, zero-V_T devices. These devices will offer some extra headroom in circuits [11], but as we will observe in Section 1.1, the main design challenge of enabling circuit operation with reasonable voltage swing, remains unaddressed with low-V_T or native zero-V_T devices. Low-V_T devices require an extra mask and extra semiconductor processing steps, resulting in an increase in cost and turn-around time. This cannot be justified when the analog interfaces occupy only 5-30% of the die area on large system-on-chip (SoC) circuits. Native zero-V_T devices do not require any extra mask or processing steps, but are typically less characterized and modeled and sometimes have less reproducible characteristics.

Another technology modification is to use floating-gate devices [12]. Several low voltage circuit design techniques using floating-gate devices have been implemented and discussed in [13]. A small charge is programmed onto a floating-gate capacitor and this can reduce the effective threshold voltage of the composite floating-gate device. These circuit techniques, however, raise reliability concerns since the internal gate voltages might be beyond the supply rails and could cause additional device stress.

Several circuit techniques have been proposed that allow circuit design at low voltages, without using floating-gate devices, special low V_T devices, or native zero-V_T devices. Back-gate or body driven circuits have been proposed in [14–16]. [17] demonstrates input level-shifting techniques, allowing for ultra-low power supply voltages. Some of these techniques have been incorporated in this work. Rail-to-rail input stages [18, 19], and multi-stage amplifiers with nested Miller compensation [18, 20, 21] have been proposed for low voltage circuits. Analog circuit blocks and amplifiers have been designed, operating at 1 V [16, 22, 23] and down to 0.9 V [24]. Radio-frequency (RF) circuits have been demonstrated down to 0.5 V [25, 26].

In the design of ultra-low voltage discrete-time circuits, an additional challenge, beyond the design of the basic amplifier, is the implementation of a switching scheme. For switched-capacitor circuit operation at low power supply voltages, clock voltage boosting techniques [27, 28] have been used to double the clock swing and drive switches. This technique cannot be used if the thin oxide limits the permissible clock voltage, which is the case in sub-micron low voltage CMOS circuits. The idea of bootstrapping the clock voltages had been used earlier in the context of accurate and linear sampling of a continuous time signal [29]. This technique has been extended to enable low voltage operation, as shown in [30–32]. However, this method imposes a higher voltage glitch across the thin gate oxide before the inversion takes place under the gate and a channel forms in the MOS switch. Another idea that extends on the clock-bootstrapping techniques is to have an integrator structure that uses a floating reference voltage [33]. The switched-opamp technique [34] and opamp-reset circuits [35] have been used in several discrete-time designs [36–41]. An alternative switched-RC technique was implemented in [42].

In this work true low voltage circuit design techniques at a power supply voltage as low as 0.5 V, in standard CMOS, have been developed and demonstrated. Voltages at all nodes in the circuit are within the power supply rails at all times. No technology modifications and special devices were used in the core circuits. In Section 1.1, we will take a close look at the circuit design challenges at ultra-low power supply voltages. This will be followed by a discussion, in Section 1.2, on the various design opportunities that open up at ultra-low power supplies.

1.1 Low-voltage analog circuit design challenges

With a supply voltage of 0.5 V and a V_T of at least 0.2 V (as foreseen by the roadmap, see Fig. 1.1(a)), devices in practical circuits can be expected to have a maximum $|V_{GS}| - |V_T|$ of 0.2 V, i.e, the devices will be at the edge of strong inversion.

For applications requiring high bandwidths or high clock and sampling rates, the MOS devices in the circuit are biased in the strong inversion region, i.e, for an nMOS device, $V_{GS} - V_T \geq 0.2$ V [43]. As long as $V_{DS} \geq V_{DS,sat}$ where $V_{DS,sat} = \frac{V_{GS} - V_T}{\alpha}$, ($\alpha$ is typically between 1 and 1.5), the device is in saturation, or in other words, acts as a voltage controlled current source.

At the edge of strong inversion, a good estimate of $V_{DS,sat}$ is about 0.15 V. The devices can also be operated in weak inversion, with a reduced V_{GS}, i.e, for an nMOS device, $V_{GS} - V_T \leq 0$ V. Weak inversion operation results in a much higher transconductance/current (g_m/I) efficiency and is useful for low power applications [44]. However the bandwidth offered by the use of a device in weak inversion is considerably less. In weak inversion, the drain-source current for an nMOS device, I_{DS} is given by [43](pp. 172):

$$I_{DS} = W/L \cdot \hat{I}(V_{GB}) e^{-V_{SB}/\phi_t} \left(1 - e^{-V_{DS}/\phi_t}\right) \tag{1.1}$$

where ϕ_t is kT/q, (k is Boltzman's constant, T is the absolute temperature, q is the negative charge of an electron) $\hat{I}(V_{GB})$ is a current that is a function of V_{GB}, and W, L are the width and length of the transistor. For a given V_{SB}, there is a smaller than 1% variation in I_{DS} for V_{DS} greater than $4\phi_t$ to $5\phi_t$. As a result, the minimal V_{DS} to maintain the device in saturation is 0.1 V to 0.125 V at room temperature. So in any region of inversion (edge of strong inversion, weak and in-between, moderate) a drain-source voltage of 0.15 V is sufficient to maintain an nMOS device in saturation. Fig. 1.2 shows the I_{DS} measured in an nMOS device of length 0.72 μm, width 285 μm, as a function of V_{DS} for different V_{GS}. For this device, the minimum V_{DS} required to maintain saturation in weak inversion, is about 0.1 V. It turns out that this is true for any device, independent of process, V_T, and dimensions, as long as (1.1) is approximately valid for the device in weak inversion.

The gate-source bias of the device, V_{GS}, on the other hand, directly controls the inversion of the device and the bias current through the device. To bias a device in a

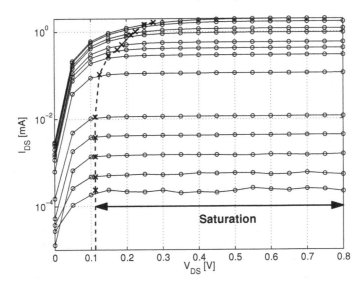

Fig. 1.2: I_{DS} as a function of V_{DS} for different V_{GS}, measured in an nMOS device of $L = 0.72\ \mu$m, $W = 285\ \mu$m, fabricated in a $0.18\ \mu$m process. In weak and moderate inversion, the minimum V_{DS} required to maintain the device in saturation (smaller than 1% variation in I_{DS}) is 0.1 V.

given region of inversion with a given bias current, the required V_{GS} depends on the V_T.

The simplest arrangement to get amplification from a MOS device is the common source configuration with a single-device active load, as shown in Fig. 1.3(a). For a V_{DD} of 0.5 V, and a $V_{DS,sat}$ of 0.15 V, at the optimal output (drain) bias voltage of $V_{DD}/2$ or 0.25 V, the maximum peak-peak output swing, $V_{out,pp}$, is $V_{DD} - 2V_{DS,sat}$ or 0.2 V. The signal swing allowed at the input (gate) depends on the V_T of the device, and is likely to be small. However, as long as there is a large gain between input and output, this is not a strong limitation.

With a 0.5 V supply, in a common drain configuration as shown in Fig. 1.3(b), the output can swing sufficiently. However, because there is no gain from the input to the output, and because of stacking of the output bias and V_{GS} bias, the DC bias voltage at the gate is very high. This limits the available swing at the gate.

In a common gate configuration the input signal, as shown in Fig. 1.3(c), output signal and $3V_{DS,sat}$ are stacked. Even if we assume a large voltage gain for the stage, the available output swing is too small for most applications. A common gate stage (or folded cascode) can be embedded in an amplifier if followed by sufficient gain so that no significant swings are needed at the common gate output. Similarly, cascode topologies with all devices in saturation are excluded at 0.5 V since they require a stack of the output swing and $4V_{DS,sat}$ (about 0.6 V).

Of the basic transistor configurations only the common source configuration has the potential to operate at supply voltages of 0.5 V. It is again important to remark that this limitation stems from the required $V_{DS,sat}$ of about 0.15 V and is independent of the value of V_T.

The two major challenges in designing an amplifier at a power supply voltage of 0.5 V are therefore,

- A minimum drain-source voltage of $4\phi_t$ to $5\phi_t$ is required to maintain the device in saturation. Given the low power supply voltage, this limits the voltage swing at the drain of the device.
- $V_{GS} - V_T$ sets the inversion level of the device. For high speed applications, the device has to be maintained in strong inversion. However, the gate-source voltage is limited by the power supply.

Fully differential circuits are widely used in contemporary analog integrated circuits [45–47] due to their larger signal swing and better supply and substrate interference immunity. At 0.5 V, we have to rely on those properties and fully differential topologies are essential. The correct operation of fully differential topologies relies on the availability of good common-mode rejection. Not only does the differential gain need to be significantly larger than the common-mode gain, the common-mode gain also needs to be sufficiently less than 1, due to the possible presence of positive feedback loops in the common-mode signal paths.

Let us now observe the difficulties with a practical OTA design at a supply voltage of 0.5 V. Fig. 1.4 is a conceptual schematic of a standard, two stage, folded-cascode transconductance amplifier. For a differential input signal, the differential signal current is proportional to the g_m of the input differential pair. This differential current generates a differential voltage at nodes 1 and 2, that is proportional to the output impedance seen at these nodes. The differential voltage is further amplified by the output stage of the amplifier. For a common-mode input signal, the common-mode output current is determined by the conductance of the tail current source, and is expected to be small. The small response to common-mode signals, in combination with a wide-band common-mode feedback (CMFB) provide a very small common-mode gain and strong common-mode rejection [45–47].

Now, we proceed to selecting the bias voltages for the devices in the folded-cascode amplifier for optimal operation, at a power supply voltage of 0.5 V. For this discussion, let us assume that all devices have a $|V_T|$ of 0.15 V, the minimum gate overdrive required is 0.2 V, and that the minimum V_{DS} required to operate the devices in saturation is 0.15 V. Let us further assume that the devices indicated as current sources also require 0.15 V across them to maintain a large output resistance. The required bias voltages are indicated in Fig. 1.4. At the nodes 1 and 2 in Fig. 1.4, the minimum voltage allowed by devices M2A and M2B is 0.3 V. However, the maximum voltage allowed by devices M3A and M3B is 0.2 V. Moreover, the voltage required to maintain devices M4A and M4B in strong inversion is 0.35 V. As a result, strong inversion operation of this folded-cascode amplifier is impractical with a 0.5 V supply voltage, even at such a low V_T of 0.15 V. The stack of four devices in

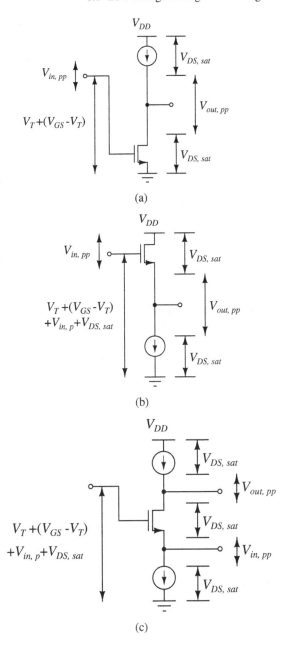

Fig. 1.3: Basic MOS device configurations: (a) Common source, (b) Common drain, (c) Common gate.

the folded-cascode amplifier cannot be accommodated at a power supply voltage of 0.5 V.

Let us now analyze the signal and biasing ranges for an input differential pair, shown in Fig. 1.5. As a result of stacking three devices, the output swing, $V_{out,se,pp}$ is limited. Adding a second gain stage after the input stage will overcome this limitation. A significant challenge is maintaining the required input bias voltage of $V_{DD} - |V_T| - (V_{SG} - |V_T|) - V_{DS,sat}$. If the inputs can be biased at 0 V, for a 0.5 V V_{DD}, the resulting maximum allowed $|V_T|$ is about 0.15 V for strong inversion operation. Even when such a low V_T is available, in practice the inputs of the OTA will need to be above the ground supply rail to enable use with feedback. As a result, strong inversion operation of this stage is impractical with a 0.5 V supply voltage. So, for any case where V_T is greater than 0.15 V, there is a need to develop differential input structures with good common-mode rejection.

The design of wide band common-mode feedback loops is also challenging at 0.5 V. The output common-mode voltage is set to 0.25 V, or $V_{DD}/2$, for maximum output swing. This makes the design of a wide-band error amplifier difficult. Alternative local common-mode feedback techniques will be discussed in the subsequent sections.

The realization of a low output impedance is required for the implementation of an operational amplifier. The unavailability of the common-drain stage makes this difficult, and hence we are limited to the implementation of operational transconductance amplifiers (OTAs). For most on-chip applications, the loads are mainly capacitive and the load impedances can be kept sufficiently large, so that this is not a limitation. In feedback circuits, the loop gain reduces the effective output impedance.

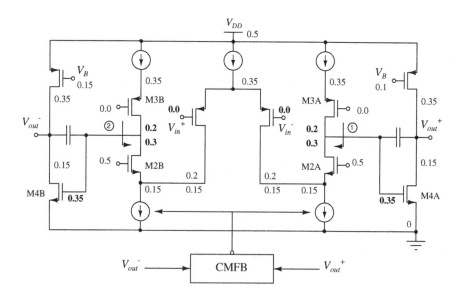

Fig. 1.4: Schematic of a two stage, folded-cascode OTA.

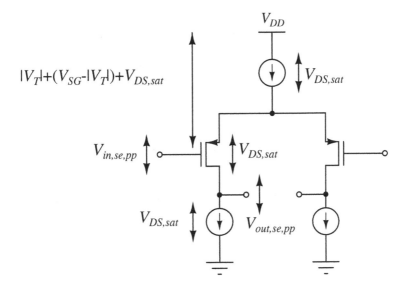

Fig. 1.5: Bias and signal ranges for an input differential pair.

1.2 Opportunities at low voltages

In addition to the gate, the channel of an MOS device can also be controlled from the body. However, this is usually not done, due to concerns of latch-up when a forward bias voltage is applied across the body-source junction.

The phenomenon of latch-up is explained in [47] (pp. 118-120). A typical CMOS circuit consists of nMOS and pMOS transistors, all located close together on the die. A simplified cross section showing one pMOS and one nMOS is shown in Fig. 1.6(a). Parasitic npn and pnp transistors are also shown. A rearrangement of the way these parasitic bipolar transistors are viewed shows a positive feedback structure in Fig. 1.6(b). If Q_2 is triggered on, with current injected into its base, it switches on Q_1. The emitter-base junction of Q_1 requires about 0.7 V to turn the device on. Q_1 being on, causes further current to flow into the base of Q_2. The p-substrate, thus, latches close to the positive supply rail, while the n-well latches close to the ground potential.

At a supply voltage of 0.5 V, even with current injected into the base of Q_2, without the voltage headroom, Q_1 will not be triggered on. As a result, the mechanism of latch-up will fail. However, there still exists a possibility of latch-up in the presence of power-supply transients.

The diminished risk of latch-up enables the use of forward biasing of the body-source junction, and in general, the use of the body terminal of any device as a fourth, controllable terminal. Forward biasing the body-source junction of a device reduces the threshold voltage [48–51].

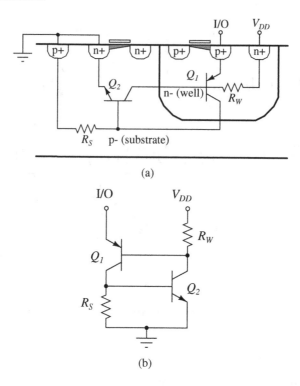

(a)

(b)

Fig. 1.6: Mechanism of latch-up: (a) Cross-section of a simple CMOS circuit with one nMOS and one pMOS (b) Rearrangement of parasitic bipolars shows a positive feedback structure.

Traditionally only the body terminal of pMOS devices could be accessed in n-well processes, but modern CMOS processes offer the availability of nMOS devices in a separate well or with buried deep n-well layers, so that their body terminal can be accessed independently, as shown in Fig. 1.7. Isolated nMOS devices in a deep n-well process (Fig. 1.7(a)) track their un-isolated counterparts in the same process, because of similar doping profiles. On the other hand, in a triple-well process (Fig. 1.7(b)), the isolated nMOS device is different in characteristic from its un-isolated counterpart, where the doping profile is markedly different. In both cases, access to the body node of a device has an area penalty. This area penalty is of little concern if the design is limited by other larger components, such as capacitors or inductors. Also, several nMOS devices which are connected to the same body potential can be grouped together in the same well to reduce the total area use. In any case, the area increase is small in comparison to the area of the digital circuits.

The effect of the body on the device V_T is explained [43] (pp. 101-103) with the help of Fig. 1.8. When the body and source are at the same potential (Fig. 1.8(a)), there is a depletion region between the n+ source and the p-substrate. The same depletion region continues, when in strong inversion, beneath the channel, where there

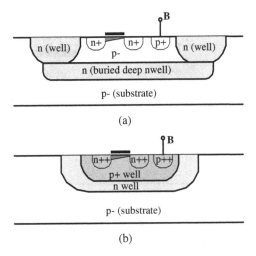

Fig. 1.7: Conceptual cross-sectional view of isolated nMOS devices: (a) with a buried deep n-well layer, (b) in a triple-well CMOS process

is an abundance of free electrons. For the same V_{GS} in strong inversion, the charge on the gate, which is the positive plate of the capacitor, remains constant. This charge has to be equal in magnitude to the total negative charge, which is the sum of the free charge in the channel and the fixed charge in the depletion region, assuming negligible interface charge. In Fig. 1.8(b), a positive body-source bias voltage decreases the depth of the depletion region. As a result, for the same V_{GS}, the charge in the channel increases, and the threshold voltage, V_T, decreases.

The V_T of an MOS device has been classically developed [43, 52] for $V_{SB} > 0$:

$$V_T = V_{T0} + \gamma \left(\sqrt{\phi_0 + V_{SB}} - \sqrt{\phi_0} \right) \tag{1.2}$$

where, V_{T0} is the value of V_T for a body-source bias of zero, γ is the body-effect coefficient for a given technology, ϕ_0 is given by:

$$\phi_0 = 2\phi_F + \Delta\phi$$

where ϕ_F is the Fermi-potential and $\Delta\phi$ is about $6kT/q$, i.e about 150 mV at room temperature. (1.2) shows that when V_{SB} on a device is positive and is increased, V_T of the device increases. (1.2) has been rearranged and extended for $V_{SB} < 0$ (or positive V_{BS}), with the help of experimental evidence [53–55].

$$V_T = V_{T0} - \gamma \left(\sqrt{\phi_0} - \sqrt{\phi_0 - V_{BS}} \right) \tag{1.3}$$

(1.3) shows that increasing the body-source junction forward-bias decreases the V_T of the device.

A pMOS device of length 0.72 μm and width 285 μm, fabricated in a 0.18 μm CMOS process, was characterized for different gate-source, drain-source and body-source voltages. The extracted threshold voltage of the device, V_T, is plotted in

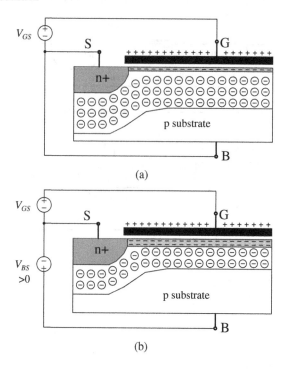

Fig. 1.8: Effect of forward-biasing the body-source junction in a weakly-inverted nMOS device: (a) Depletion and channel charges shown for a given V_{GS} and no body-source forward bias. (b) Depletion and channel charges shown for the same V_{GS} and small body-source forward bias.

Fig. 1.9 for different body-source junction bias voltages. V_T decreases as V_{BS} increases, which is also seen in (1.3). In [56–58], we typically use a forward bias which results in a reduction of the V_T.

Figure 1.10 depicts simulations of device threshold voltage, V_T, as a function of V_{BS} for different technology nodes. The projection on the 65 nm process is based on predictive technology models obtained through BPTM [59], which is provided by the Device Group at UC Berkeley. A BSIM3 model for the 130 nm process, based on wafer measurements, was obtained through MOSIS [60].

Forward body bias has been used in digital applications to tune the V_T so that a more consistent circuit performance over process and temperature and thus a higher yield is obtained [48, 49, 61, 62]. Interestingly, a low voltage swapped-body-bias (LVSB) design style has been proposed [63], where the body of the nMOS is tied to the positive supply and the body of the pMOS is tied to the negative supply. High speed or low power consumption is obtained and correct functionality for an operating temperature up to 75^0C has been demonstrated [63].

The availability of the body terminal thus offers two opportunities. The signal can be applied to the body (back-gate) of the device [14–16, 56], whereas the gate is

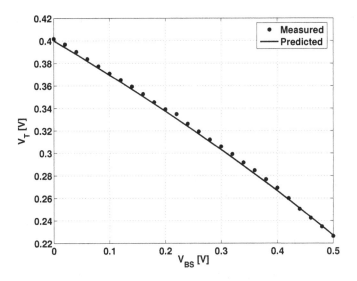

Fig. 1.9: Measured device V_T (*) as a function of V_{BS}. The characterized device on a 0.18 μm process, has $W = 285$ μm, $L = 0.72$ μm, 150 fingers. The measured curve is modeled (–) using (1.3); $V_{T0} = 0.4$ V, $\phi_0 = 1.0$ V, and $\gamma = 0.6$ $\sqrt{\mathrm{V}}$.

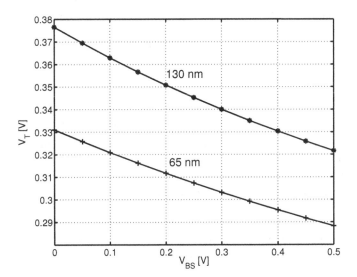

Fig. 1.10: Simulations of device V_T as a function of V_{BS} for advanced technology generations. A predictive model for devices in 65 nm was obtained through BPTM [59]. The model for the 130 nm process is based on wafer measurements, obtained through MOSIS [60].

used to bias the device; or, when we apply the signal to the gate, we can use the body (back-gate) to control the bias of the device [57]. Both techniques will be illustrated in the subsequent chapters.

1.3 Organization of the book

In Chapter 2, the design of a 0.5 V fully differential body-input OTA (Operational Transconductance Amplifier) [56] and the design of a gate-input OTA [57] are discussed. Both of these designs are for a 0.18 μm CMOS process with standard devices with a $|V_T|$ close to 0.5 V. Robust biasing techniques that maintain circuit functionality and maximize swing over process, temperature and power supply voltage variations are developed. The two OTA design techniques are compared using measurements of prototypes on a 0.18 μm CMOS process.

In Chapter 3, a low frequency, weak inversion, 0.5 V MOS varactor (variable capacitor) for use in active varactor-R filters, is proposed and analyzed [64]. A prototype notch filter using the proposed varactor and other discrete components was developed to demonstrate the capacitance variation.

The design of a filter with an on-chip PLL for tuning [57], is demonstrated in Chapter 4. This filter uses the gate-input OTAs, along with robust biasing circuitry, and the weak inversion varactors as building blocks. The prototype chip has been extensively characterized and experimental results for different chips, different supply voltages and different temperatures are presented. The 5th-order low-pass elliptic filter has a 135 kHz cut-off frequency, 1.1 mW of power consumption and 57 dB of dynamic range at a 0.5 V power supply. Nominal operation is observed from 5°C to 85°C.

This is followed, in Chapter 5, with the design of a 0.5 V track-and-hold circuit. The designed prototype is implemented on the CMOS part of a 0.25 μm BiCMOS technology. The track-and-hold circuit achieves a 60 dB of SNDR (Signal to Noise and Distortion Ratio) at a sampling frequency of 1 MHz, with a measured power consumption of 300 μW from a 0.5 V power supply voltage.

In Chapter 6, a 0.5 V third-order fully-differential continuous-time $\Sigma\Delta$ modulator is demonstrated. Active-RC implementation of the loop filter is chosen due to the availability of OTAs as presented in Chapter 2. Apart from the OTA, the modulator requires a comparator and a feedback DAC (digital-to-analog converter) with return-to-zero signaling. The design challenges of these two blocks at 0.5V are discussed, and a 0.5 V body-input gate-clocked latched comparator and a return-to-open split buffer DAC architecture, are presented as solutions. A modulator architecture tolerable to variations of RC products and DAC timing is also presented. Fabricated on a 0.18 μm CMOS process, the modulator achieves a peak SNDR of 74 dB in a 25 kHz bandwidth with an over-sampling ratio of 64. The modulator core occupies an area of 0.6 mm^2 and consumes 300 μW.

In Chapter 7, a 900 MHz RF receiver front-end is presented. The prototype chip includes an LNA, a quadrature downconversion mixer and associated LO buffers. All circuits operate from a 0.5 V supply. The circuit is designed in 0.18 μm CMOS.

It achieves a conversion gain of 12 dB, an IIP3 of -14 dBm and a noise figure of 9 dB. The circuit, including the LO buffers, dissipates 7.4 mW and occupies area of 0.43 mm^2.

In keeping with the motivation of the work, all of the designed circuits are true low voltage circuits, i.e, voltages at all nodes in the circuits are biased within the power supply rails. Standard CMOS devices have been used in the designs. For circuits on the 0.18 μm process, the devices have a $|V_T|$ of about 0.5 V. The devices used for the RF receiver front-end have a $|V_T|$ of 0.2 V. For circuits on the 0.25 μm process, the nMOS device has a V_T of about 0.6 V while the pMOS device has a $|V_T|$ of about 0.5 V. Future CMOS technologies are expected to offer much faster transistors (Fig. 1.1(b)) that have a lower threshold voltage compared to the transistors in the used processes. The techniques demonstrated in this work are expected to yield much higher performance once these processes become available [1].

Issues that remain outstanding include the use of native or zero-V_T devices in the circuits, circuit topologies for voltage references, envelope detectors, mixers, switched-capacitor and switched-opamp circuits. These could be developed with the presented framework as a starting point.

The design techniques have explored the use of the body of the CMOS devices as a fourth terminal. This additional handle on the devices was used in two ways – e.g, in the body-input OTA, the body of the device was used for applying the signal, while in the gate-input OTA, the body of the device was used for biasing and control. Circuit topologies were introduced that have sufficient common-mode rejection, which enable fully differential operation at ultra-low power supply voltages. With the building blocks demonstrated in this work, a variety of integrated analog and mixed-signal CMOS systems can be developed at low power supply voltages. Feedback topologies for the bias and gain-enhancement circuits can be used in a variety of applications, including digital circuits and applications at higher power supply voltages. It is the hope of the authors that the work presented will be used by designers as a starting point for broader research in ultra-low voltage analog circuit design.

2

Fully Differential Operational Transconductance Amplifiers (OTAs)[1]

An amplifier is often a fundamental building block of an analog circuit. Fully differential circuits are widely used due to their large available signal swing, and superior supply and substrate interference immunity. In this chapter we will develop the design of fully differential amplifiers, which will be primary building blocks for subsequent designs at a 0.5 V supply voltage. As discussed earlier in Section 1.1, the realization of a low output impedance is required for the implementation of an operational amplifier. The unavailability of the common-drain stage, as seen in Fig. 1.3(b), makes this difficult, and hence we will only discuss the implementation of operational transconductance amplifiers (OTAs).

The design of a body-input OTA and that of a gate-input OTA will be discussed in Sections 2.1 and 2.2 respectively. Robust biasing techniques that maximize the output swing and maintain circuit functionality, are discussed in Section 2.3. The designs are implemented in a 0.18 μm CMOS process and characterization results are presented in Section 2.4. This is followed in Section 2.5 with a discussion of the two design techniques and their advantages and disadvantages.

Forward biasing of the body-source junction, as discussed in 1.2, has been applied in low voltage digital circuits [49, 50, 61, 63] and it is applied here to lower the V_T of the transistors. In the context of 0.5 V operation, the risk of forward biasing the junctions is minimized since parasitic bipolar devices cannot be activated even when the full power supply is used as forward bias, provided that supply transient over-voltages are adequately kept under control (see Section 1.2).

Figure 2.1 shows the profile of V_T as a function of the body-source forward-bias voltage, and of the device length, for a 0.18 μm CMOS process. We can typically apply a forward bias of about 250 mV which results in a reduction of V_T by about 50 mV. V_T decreases with increasing device length. To minimize static leakage in digital circuits, modern processes are engineered, with extra halo implants, so that the threshold voltage is increased, and is the maximum for minimum channel length

---

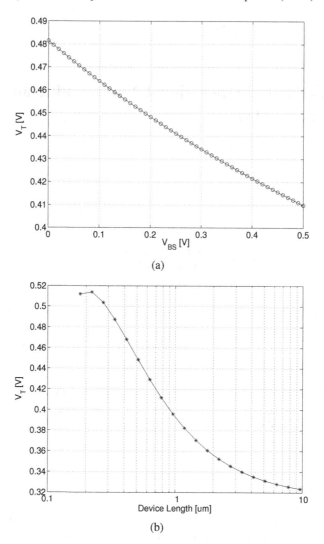

Fig. 2.1: Simulations of device threshold voltage, V_T, in a 0.18 μm CMOS process, (a) as a function of V_{BS}, the body-source forward-bias voltage, for a 0.36 μm long device, (b) as a function of the length of the device. Note that the process chosen in these device simulations is different from that of the devices measured in Fig. 1.9.

devices. To limit the effect of this reverse short-channel effect [43], and to get lower V_T devices for use in 0.5 V circuits, a larger channel length can be used.

2.1 Body-input OTA

Body-input operational amplifiers with a single-ended output have been investigated for low voltage applications down to 0.7 V [14,16,22–24]. In order to operate a MOS transistor at or near moderate inversion, a large voltage can be applied as a gate bias, and the signal can be applied to the body of the device. A very low voltage basic gain stage is shown in Fig. 2.2. The two inputs are at the bodies of pMOS transistors M_{1A} and M_{1B}, and their g_{mb} provides the input transconductance. For an input common-mode voltage of $V_{DD}/2$ (0.25 V), the resulting small body-source forward bias lowers the V_T and further increases the inversion level. Operation near the weak-moderate inversion boundary is preferred, in order to attain a relatively large body transconductance, g_{mb}. M_{1A} and M_{1B} are loaded by the nMOS transistors M_{2A} and M_{2B}, which act as current sources. The body inputs of M_{3A} and M_{3B} form a cross-coupled pair that adds a negative resistance to the output and boosts the differential DC gain. Resistors R_A and R_B detect the output common-mode voltage which is fed back to the gates of the pMOS devices M_{1A}, M_{1B}, M_{3A} and M_{3B} for common-mode rejection. A DC level shift between the output common-mode voltage at 0.25 V and the gate bias at 0.1 V is created by pulling a small current through R_A and R_B with M_4.

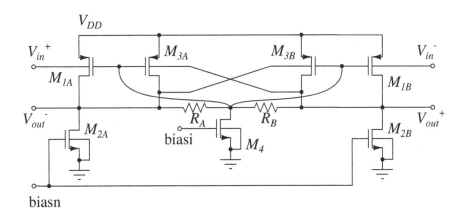

Fig. 2.2: Fully differential body-input gain stage with local common-mode feedback.

In the following g_{mb_N} is the body transconductance of M_N, g_{m_N} is the gate transconductance of M_N and g_{ds_N} is the drain-source conductance of M_N. The small-signal equivalent circuit of the body-input gain stage of Fig. 2.2, is shown in Fig. 2.3. The differential DC gain is:

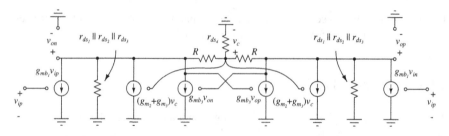

Fig. 2.3: Small-signal equivalent circuit of a body-input OTA stage.

$$A_{\text{diff}} = \frac{g_{mb_1}}{g_{ds_1} + g_{ds_3} + g_{ds_2} + 1/R - g_{mb_3}} \tag{2.1}$$

The common-mode DC gain is given by:

$$A_{\text{cm}} = \frac{g_{mb_1}}{g_{ds_1} + g_{ds_3} + g_{ds_2} + g_{mb_3} + g_{m_1} + g_{m_3}} \tag{2.2}$$

The common-mode signal is strongly suppressed as a result of g_{m_1} being larger than g_{mb_1} and is intrinsically less than 1. In our design, A_{diff} was 24 dB per stage, and A_{cm} was -14 dB per stage.

The input-referred white noise spectral density [V^2/Hz] is given by:

$$8kT \cdot \frac{2}{3} \cdot \frac{1}{g_{mb_1}} \left(\frac{g_{m_1}}{g_{mb_1}} + \frac{g_{m_2}}{g_{mb_1}} + \frac{g_{m_3}}{g_{mb_1}} \right) + \frac{2 \cdot 4kT}{g_{mb_1}^2 R} \tag{2.3}$$

The input-referred noise is intrinsically large because the body transconductance, g_{mb} is small in comparison to the gate transconductance, g_m.

By cascading two identical gain blocks, a two stage OTA is obtained as shown in Fig. 2.4. The amplifier is stabilized by adding Miller compensation capacitors C_C with series resistors R_C to move the right-half plane zero to the left-half plane. The complete schematic of the 0.5 V fully differential OTA is shown in Fig. 2.5. The frequency response has a gain-bandwidth product approximately given by $g_{mb_1}/(2\pi C_C)$, where g_{mb_1} is the body transconductance of the input transistors of the first stage; the second pole frequency is approximately at $g'_{mb_1}/(2\pi C_L)$, where g'_{mb_1} is the body transconductance of the input transistors of the second stage and C_L is the load capacitance. In applications with multiple OTA stages, the input pMOS n-well to p-substrate parasitic capacitance presents a load to the previous OTA stage. However, this capacitance will form a part of the total compensation capacitance, which will be dominated by the compensation capacitors themselves.

M_{1A}, M_{1B} were sized for a gain-bandwidth product of 2.5 MHz for a bias current of 40 μA and a load of 20 pF on each output. M_{3A} and M_{3B} were sized conservatively for a 9 dB gain improvement. The design had a nominal power dissipation of 100 μW. The transistor widths and lengths, and the values of the passive elements are shown in Table 2.1. Throughout the design a larger than minimum channel length was used to limit the reverse short-channel effect and to take advantage of lower V_T

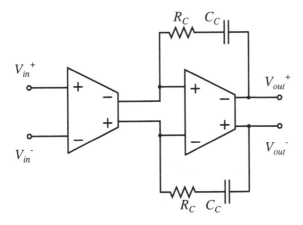

Fig. 2.4: Two-stage OTA compensated with Miller capacitors and series resistors.

Table 2.1: Transistor sizes and element values for the body-input OTA (Fig. 2.5)

First stage			Second stage		
Transistors	W [μm]	L [μm]	Transistors	W [μm]	L [μm]
M_{1A}, M_{1B}	240	0.5	M'_{1A}, M'_{1B}	240	0.5
M_{2A}, M_{2B}	75	0.5	M'_{2A}, M'_{2B}	75	0.5
M_{3A}, M_{3B}	40	0.5	M'_{3A}, M'_{3B}	40	0.5
M_4	3.5	1.0	M'_4	3.5	1.0
Resistors and Capacitors					
R_A, R_B	100 kΩ		R'_A, R'_B	100 kΩ	
			R_C	6.5 kΩ	
			C_C	6 pF	

devices, as shown in Fig. 2.1(b). Larger channel lengths will require, for the same performance, larger widths. This in turn will result in larger capacitances that the circuit will need to drive. A channel length of 0.5 μm was chosen for all devices in this circuit as a compromise in the trade-off between lower V_T, larger area requirements, and larger parasitic capacitance. Currents of 40 μA and 4 μA were input through the nodes "biasn" and "biasi" as in Fig. 2.5. The current mirrors reflect these currents through M_{2A}, M_{2B} and M_4 respectively.

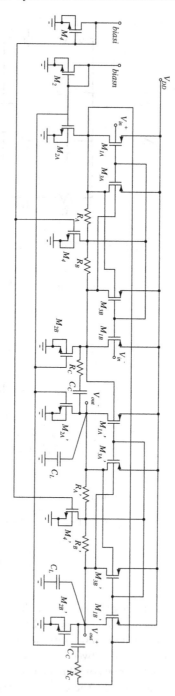

Fig. 2.5: Complete schematic of the two-stage fully differential body-input OTA with Miller compensation.

A prototype was fabricated in a CMOS 0.18 μm mixed-signal process. The total area was 130 μm x 200 μm. High-resistivity poly resistors were used for the 100 kΩ resistors. MIM capacitors were used for the compensation capacitors. The chip micro-graph is shown in Fig. 2.6(a) (there is exhaustive metal fill over the entire circuit), with the corresponding layout in Fig. 2.6(b). The compensation capacitors can be seen on the two sides of the figures.

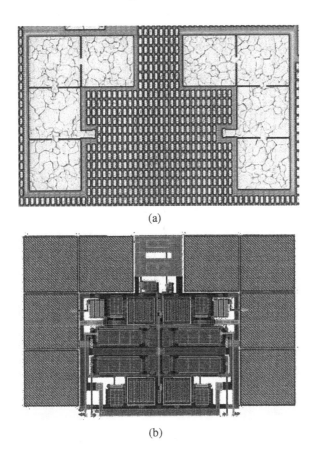

(a)

(b)

Fig. 2.6: (a) Chip micro-graph of the body-input OTA. (b) Body-input OTA layout.

2.2 Gate-input OTA

The second OTA developed in this work uses gate inputs with the body terminal for adaptive biasing. To access the body terminals of nMOS devices, triple-well devices were used.

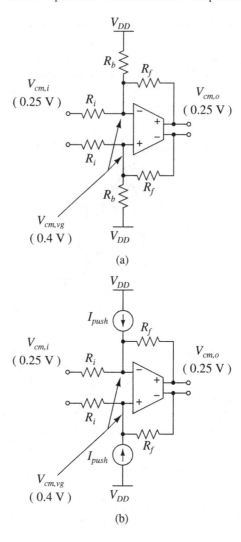

Fig. 2.7: Setting input and output common-mode voltages of the gate-input OTA, (a) using resistors, R_b, (b) using current sources, I_{push}

Figure 2.7 shows amplifier configurations using resistive feedback around an OTA. For this discussion let us assume that the input of the OTA consists of nMOS devices. A similar argument can be made for input pMOS devices. For a maximal output signal swing, the output common-mode level, $V_{cm,o}$, is typically set to $V_{DD}/2$. Since the input of each stage is driven from a similar stage, the input common-mode level, $V_{cm,i}$, is also $V_{DD}/2$. In order to turn the input devices of the OTA on, the virtual ground common level, $V_{cm,vg}$, needs to be set as high as possible. Such a common level can be maintained by a resistor, R_b, shown in Fig. 2.7, without affecting the overall gain of the circuit [4, 17], as long as:

$$R_f \ll A \cdot (R_i \parallel R_f \parallel R_b)$$

where A is the open-loop DC gain of the amplifier. For V_{DD} of 0.5 V and $V_{cm,i}$, $V_{cm,o}$ of 0.25 V, to push $V_{cm,vg}$ to 0.4 V, R_b is given by:

$$R_b = 2/3 \cdot (R_i \parallel R_f)$$

Thus, a gate-input low voltage OTA can be used with the signal common-mode voltage at $V_{DD}/2$, and yet maintain the OTA input devices in moderate inversion.

Current sources could be used in place of the resistors R_b, as shown in Fig. 2.7(b) [65]. A DC current, I_{push} pushed into the virtual ground nodes will maintain a DC voltage drop across the feedback resistor, R_f, as well as the input resistor R_i. The current, I_{push} can be adjusted to maintain $V_{cm,vg}$ at the desired DC value, and is given by:

$$I_{push} = (V_{cm,vg} - V_{cm,o})/(R1 \parallel R2)$$

However, this method injects 1/f noise (as well as white noise) from the current source directly to the input of the OTA.

In the basic differential amplifier in Fig. 2.8(a) the input differential pair, M_{1A} and M_{1B} and the active loads M_{2A} and M_{2B}, amplify the differential input voltage. The resistors R_{1A}, R_{1B}, provide common-mode feedback through the active load. A level-shifting current I_L develops a 0.15 V drop across R_{1A} and R_{1B} to maintain V_x around 0.1 V so that M_{2A} and M_{2B} operate in moderate inversion. The bodies of M_{2A} and M_{2B} are connected to the gates to further reduce their V_T. Using the biasing arrangement shown in Fig. 2.7 the input common-mode voltage is maintained around 0.4 V. To lower the V_T, the body of the input devices M_{1A} and M_{1B} is forward biased.

The ratio of the transconductance of M_{1A} and M_{1B} to the total transconductance of M_{2A} and M_{2B} sets the common-mode gain. In the process used, the pMOS transconductance is not sufficiently large compared to the nMOS transconductance to obtain a low common-mode gain. Therefore a common-mode feed-forward cancellation path [66], [67] is added, as shown in Fig. 2.8(b), through M_{5A}, M_{5B}, M_6 and M_{3A}, M_{3B}. In M_{3A}, M_{3B} and M_6, the gate and the body are connected to each other to obtain a forward bias across the body-source junctions; this pushes these devices towards moderate inversion.

The small-signal equivalent circuit of a gain stage of the OTA of Fig. 2.8(b), is shown in Fig. 2.10. The overall DC small-signal differential gain is:

$$A_{\text{diff}} = \frac{g_{m_1}}{g_{ds_1} + g_{ds_2} + g_{ds_3} + g_{ds_4} + 1/R - g_{m_4}} \tag{2.4}$$

The common-mode small-signal gain is given by:

$$A_{\text{cm}} = \frac{g_{m_1} - 2g_{m_5}\frac{g_{m_3}+g_{mb_3}}{g_{m_6}+g_{mb_6}}}{g_{m_4} + g_{m_2} + g_{mb_2}} \tag{2.5}$$

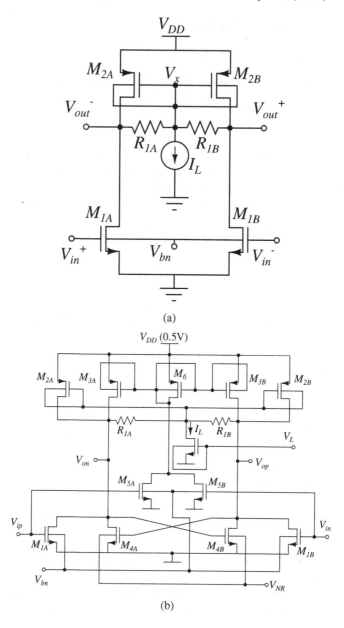

(a)

(b)

Fig. 2.8: Circuit development of the gate-input OTA. (a) Basic configuration. (b) Schematic of one stage of the gate-input OTA.

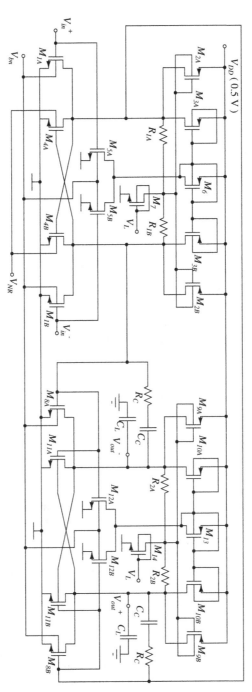

Fig. 2.9: Two-stage, fully differential, gate-input OTA, with Miller compensation.

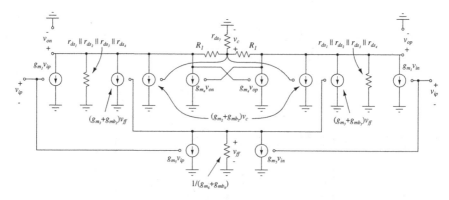

Fig. 2.10: Small-signal equivalent circuit of a single gain stage of the gate-input OTA.

If the W/L ratios of M_1, M_3, M_5, M_6 are such that $M_1/M_5 = M_3/M_6$, then the common-mode gain will be zero. In this design, $M_3/M_6 = 0.25 \cdot M_1/M_5$ so that 6 dB of rejection through the common-mode feed-forward path is obtained. In our design, A_{diff} was 25 dB per stage, and A_{cm} was -10 dB per stage.

The input-referred white noise spectral density [V²/Hz] is given by:

$$8kT \cdot \frac{2}{3} \cdot \frac{1}{g_{m_1}}\left(1 + \frac{g_{m2}}{g_{m1}} + \frac{g_{m3}}{g_{m1}} + \frac{g_{m4}}{g_{m1}}\right) + \frac{2 \cdot 4kT}{g_{m_1}^2 R} \qquad (2.6)$$

Compared to the input-referred noise in the body-input OTA (2.3), the noise here is substantially less because of g_{m_1} in the denominator.

To obtain a DC gain greater than 50 dB, two gain stages are cascaded to form a two-stage operational transconductance amplifier, as shown earlier in Fig. 2.4. The complete schematic of the two-stage fully differential gate-input OTA is shown in Fig. 2.9. By increasing the DC drop across R_{1A} and R_{1B} in the first stage to 0.3 V, the output common mode voltage of the first stage is set to about 0.4 V which assures proper biasing of the input devices of the second stage; a DC drop of 0.15 V across R_{2A} and R_{2B} sets the output common-mode level to 0.25 V. The differential gain is further enhanced with a cross-coupled pair, M_{4A}, M_{4B}, in the first stage which acts as a negative conductance and decreases the output conductance. As an added benefit, the common-mode gain is also further reduced. The body of this cross-coupled pair is set through an on-chip automatically controlled bias voltage, V_{NR}. A similar pair, M_{11A} and M_{11B}, is added in the second stage, only its body terminal can operate from the low common voltage at the output and its body transconductance is used to provide a negative conductance; its gate transconductance is in parallel with the input transconductance. M_{11A} and M_{11B} are sized conservatively.

The OTA is stabilized through the Miller capacitors C_C across the second stage. The gain-bandwidth product is approximately $g_{m_1}/(2\pi C_C)$ and the second pole frequency of the amplifier is approximately at $g_{m_8}/(2\pi C_L)$ where C_L is the single-ended load capacitance. The series resistor R_C moves the zero introduced by C_C from the right half plane to the left half plane.

Table 2.2: Component sizes and values for the gate-input operational transconductance amplifier (Fig. 2.9).

First stage			Second stage		
Transistors	W [μm]	L [μm]	Transistors	W [μm]	L [μm]
M_{1A}, M_{1B}	72	0.36	M_{8A}, M_{8B}	100	0.36
M_{2A}, M_{2B}	270	0.36	M_{9A}, M_{9B}	240	0.36
M_{3A}, M_{3B}	270	0.36	M_{10A}, M_{10B}	240	0.36
M_{4A}, M_{4B}	9	0.36	M_{11A}, M_{11B}	25	0.36
M_{5A}, M_{5B}	9	0.36	M_{12A}, M_{12A}	10	0.36
M_6	67.5	0.36	M_{13}	48	0.36
M_7	64	0.36	M_{14}	32	0.36
Resistors and Capacitors					
R_{1A}, R_{1B}	25 kΩ		R_{2A}, R_{2B} R_C C_C	20 kΩ 2 kΩ 3 pF	

The OTA is designed in a 0.18 μm triple-well CMOS process. The lengths and widths of the transistors as well as other component values are given in Table 2.2. Throughout the design a larger than minimum channel length was used to limit the reverse short-channel effect and to take advantage of lower V_T devices, as shown in Fig. 2.1(b). A channel length of 0.36 μm was chosen for all devices in this circuit as a compromise in the trade-off between lower V_T, larger area requirements, and larger parasitic capacitances.

Figure 2.11(a) shows the chip micro-graph of a gate-input OTA and Fig. 2.11(b) shows the detailed layout. High-resistivity resistors are used for the common-mode feedback resistors, and MIM-capacitors are used for the Miller compensation capacitors. Triple-well devices are used for all nMOS devices, so that there is access to the nMOS bodies.

2.3 On-chip biasing circuits for the gate-input OTA

In this section we discuss how the gate-input OTA can be biased reliably over process, supply voltage and temperature. Similar techniques can be adopted to bias the

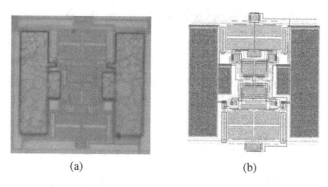

Fig. 2.11: (a) Chip micro-graph of a gate-input OTA. (b) Gate-input OTA layout.

body-input OTA. For proper operation, the gate-input OTA in Fig. 2.9 requires three biasing voltages: V_{bn}, the voltage to bias the bodies of device pairs M_1, M_5, M_8, M_{12}; V_L, to bias the level-shifting current source I_L and maintain a process and temperature independent voltage drop across R_{1A}, R_{1B}; and V_{NR}, the voltage to bias the bodies of the cross coupled pair, M_4, in the first stage of the amplifier and set the DC gain.

2.3.1 Error amplifier

In our bias loops replica circuits in combination with active feedback loops have been used extensively. These replica circuits require an error amplifier. A carefully sized inverter can be used as an inverting error amplifier; this inverter compares the input voltage to its own switching threshold voltage and amplifies the difference. For a 0.5 V supply the amplifier's devices operate in weak inversion, but the resulting slow frequency response can be tolerated in the DC biasing loops.

In [48,49,61,62], adaptive body bias techniques have been used to optimize the delay through critical paths in digital circuits. In this work, the switching voltage of the error amplifiers is adjusted by controlling the bodies of the nMOS devices through a negative feedback arrangement with three identical stages, as shown in Fig. 2.12(a). The switching threshold voltage is set to $V_{DD}/2$ independent of variations in process and temperature as follows: if the switching threshold voltage of "ErrorAmpA" is smaller than $V_{DD}/2$, the input voltage of "ErrorAmpB" decreases, the output voltage of "ErrorAmpC" decreases, the body biasing of the nMOS devices is reduced, and the switching threshold voltage increases. Similarly, when the switching threshold voltage is greater than $V_{DD}/2$, the feedback will react and decrease the switching threshold voltage. In this case, the feedback loop accurately sets the switching threshold voltage to 0.25 V for nominal operation. The stability of the feedback loop is established through C_{ea} with a zero canceling series resistor R_{ea}. Every replica error amplifier biased from V_{amp} is now an inverting error amplifier that compares its own input to 0.25 V. The DC input-output characteristics of an inverting error amplifier are shown in Fig. 2.12(b). The amplifier has a gain-bandwidth product of 20 kHz with a current consumption of 2 μA for a load of 1 pF.

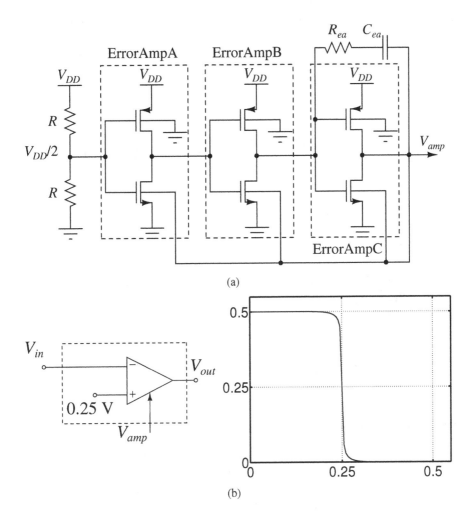

(a)

(b)

Fig. 2.12: (a) Error-amplifier biasing loop to fix the switching threshold voltage to $V_{DD}/2$. (b) DC input-output characteristics of a biased error amplifier.

2.3.2 Generating a fixed level shift

The bias current I_L in Fig. 2.8(b) creates a level shift across resistors R_{1A}, R_{1B}. This level shift is essential in maintaining the pMOS devices in moderate inversion, and in maintaining the correct common-mode level for the output of the first stage

of the OTA. A current source is designed using a single nMOS device. To increase
the inversion level of this device, the bias voltage is applied both, through the gate
and the body. A replica of this current source is used in the biasing circuit as shown
in Fig. 2.13. A current is drawn by the device which creates a voltage drop across
the resistors. V_L is generated such that V_Y is 0.25 V. The corresponding drop across
the 100 $k\Omega$ resistor R_{lc} is 0.15 V. This well defined IR voltage drop is ratio-ed and
transferred to the level shifters in Fig. 2.9 through M_7, R_{1A}, R_{2A} and M_{14}, R_{2A},
R_{2B}. A compensating capacitor, C_{CL}, is used to stabilize the feedback loop.

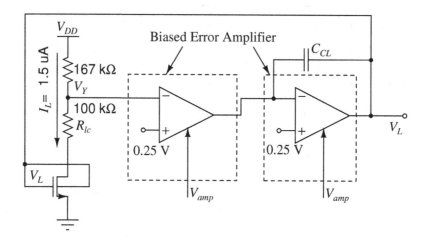

Fig. 2.13: Biasing the level-shifting current source.

2.3.3 Setting the OTA output DC common-mode voltage

The bias voltage V_{bn} in Fig. 2.8(b) adjusts the biasing level of the nMOS devices
compared to the pMOS devices and allows it to control the DC output common-mode
voltage of the OTA. As V_{bn} increases, the DC output common-mode voltage of the
amplifier decreases. To generate V_{bn}, the circuit of Fig. 2.14 was used to sense the
output common mode of a replica of the amplifier for an input common-mode voltage
of 0.4 V. In this design, 0.4 V is supplied externally, for nominal operation. The
output common-mode voltage is compared to 0.25 V and the difference is amplified
to control V_{bn} through negative feedback. A compensating capacitor, C_{bn}, is used to
stabilize the feedback loop. Note that this low bandwidth bias circuit only sets the
DC value of the output common-mode voltage and adjusts it for process, temperature
and supply voltage variations. The rejection of common-mode signals is performed
locally in each stage of each OTA.

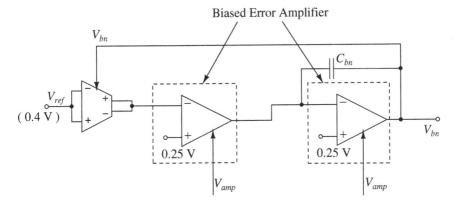

Fig. 2.14: Biasing bodies of input nMOS devices to set the OTA output common-mode voltage.

2.3.4 Gain enhancement

The cross-coupled pair of devices in Fig. 2.9, M_{4A}, M_{4B}, provides a negative resistance load to the first stage of the amplifier and enhances its gain. The amount of negative resistance can be controlled through the bodies of the two devices, by changing their V_T. If $g_{m_4} = \sum_1^4 g_{ds_i} + 1/R_1$, the OTA gain is theoretically infinite, as seen in (2.4; for a smaller g_{m_4}, the gain is positive and for a larger g_{m_4}, the gain appears to be negative. A closer investigation leads to the OTA DC transfer characteristics in Fig. 2.15. As g_{m_4} increases, the gain first increases till it becomes infinitely large and then the amplifier develops hysteresis. An OTA with hysteresis behaves as a Schmitt trigger.

To sense the onset of this behavior, an OTA-based Schmitt-trigger oscillator shown in Fig. 2.16, was designed. The oscillator oscillates at a frequency given by:

$$f_0 = \frac{1}{2RC} \cdot \frac{1}{\ln \frac{1+\beta}{1-\beta}}$$

where $\beta = V_{hyst}/V_{HL}$, with V_{hyst} the difference between the trigger voltages for the rising and falling edges and V_{HL} the difference between the high and low outputs [68]. The output of the XNOR gate decreases when oscillations are present. When the oscillator amplitude is large, V_{NR} is reduced; when the oscillator ceases to oscillate, V_{NR} is increased. In practice, the determined V_{NR} will still cause oscillations in the Schmitt-trigger oscillator, which will be so fast that the XNOR gate will be too slow to respond to. The loop guarantees that V_{hyst} is very small, so the oscillation amplitude is very small and the oscillation frequency is very high. The resulting V_{NR} is converted through a gain less than, but close to, 1, and is applied to all of our OTAs which guarantees that each OTA stage's small-signal gain is positive.

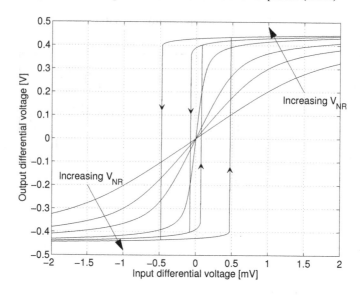

Fig. 2.15: OTA DC transfer characteristics as V_{NR} changes; hysteresis marked with arrows as g_{m_4} becomes too large.

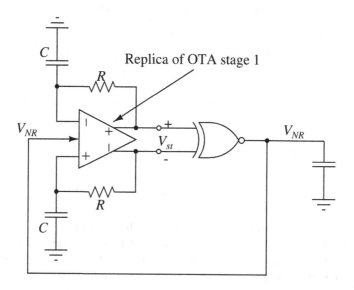

Fig. 2.16: Gain enhancement; general Schmitt trigger based oscillator and biasing technique to improve the gain of the OTA.

2.3.5 Start-up

At power-on, the error-amplifier bias circuit starts up by itself. Once V_{amp} is stable, the level-shifting current-source biasing circuit stabilizes and provides V_L. The V_{bn} biasing requires the external voltage V_{ref}, V_L and the error-amplifier bias, V_{amp}. The Schmitt trigger based oscillator starts up after this – it uses stable V_{bn} and V_L voltages. The forward-only dependencies in biasing ensure a smooth start-up.

2.4 Characterization results for the body-input and gate-input OTAs

The general measurement setup is shown in Fig. 2.17. A transformer was used to convert a single-ended signal to differential with a common-mode voltage of 0.25 V. This signal was applied to the circuit as an input. The Tektronix P6046 differential probe was used to sense the outputs of the OTA. Extensive measurements were taken using the HP3585A spectrum analyzer. To measure common-mode and power-supply rejection ratios, the setup shown in Fig. 2.18 was used. The values of various parameters, from both measurements and simulation, are shown in Table 2.3.

2.4.1 Body-input OTA measurements

The open-loop frequency response of the body-input OTA was measured at a power supply of 0.5 V, and is shown in comparison to simulated results in Fig. 2.19. Simulations of the open-loop phase response, adjusted for board parasitic capacitances, are shown compared to measurements, in Fig. 2.20. Measurements of the gain and phase response of the body-input OTA, in closed loop, are shown in Fig. 2.21. The measured output noise in the closed loop is shown in Fig. 2.22. The initial slope is a result of 1/f noise, the peaking in the noise response close to the gain-bandwidth frequency is because of the low phase margin. Measured results, in all cases, match well with simulations.

At voltages higher than 0.5 V, without changing the bias current, the input common-mode voltage was adjusted to be 0.25 V less than the power supply. As expected, the OTA worked with unchanged gain-bandwidth product and a little higher current consumption. As the supply was increased from 0.5 V to 1 V, the current consumption increased from 220 μA to 245 μA. This shows that the amplifier is robust and maintains performance over a large power supply range. As the supply voltage was decreased below 0.5 V, the bias currents were adjusted for maximum speed. At 0.4 V, the gain bandwidth product was 840 kHz with a current consumption of 66 μA and a maximum output swing of 320 mV differential peak-peak. At these low supply voltages, the gain bandwidth was largely limited by the available bias current from the biasing current sources, M_2 and M_4, for such low gate-source voltages. The total current consumption and the gain-bandwidth product are shown in Fig. 2.23 as the power supply voltage is varied.

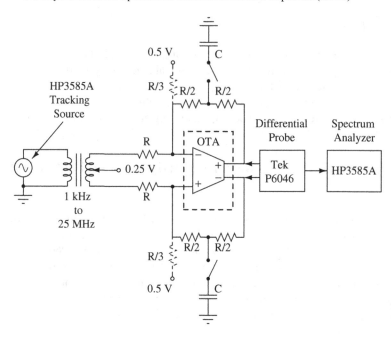

Fig. 2.17: General measurement setup for the body and gate-input OTAs. The switches are closed for open-loop measurements and are left open for closed-loop measurements. The dashed resistors are required for testing the gate-input OTA only. The input capacitance of the probe, C_L, is differential 10 pF.

2.4.2 Gate-input OTA measurements

The gate-input OTA, along with its associated bias circuits, was used in a filter, to be discussed in Chapter 4. A stand-alone gate-input OTA was included on the same chip as a test structure. In simulation, the design had a DC small-signal gain of 55 dB, a nominal unity gain bandwidth of 15 MHz and a phase margin of 60°. The measured open-loop frequency response of the OTA is shown in Fig. 2.24 for different V_{NR}. The negative resistor bias circuit automatically sets V_{NR} such that the OTA DC gain is 62 dB. These measurements were done for a load resistor of 50 kΩ. The DC gain is expected to be much higher for smaller loading.

Figure 2.25 shows the measured DC input-output transfer characteristics of the gate-input OTA. For larger V_{NR}, hysteresis develops and the gate-input OTA acts like a Schmitt trigger. The measured hysteresis is much larger than simulations (Fig. 2.15) and this can be attributed to the 50 kΩ load resistance that was used for the measurements.

Measurements and simulations of the gate-input OTA in closed loop are compared to each other in Fig. 2.26. Harmonic distortion of the OTA is measured in closed loop and is shown in Fig. 2.27. The output amplitude for a 1% total harmonic distortion is differential 712 mV peak-peak. Overall, the measured characteristics

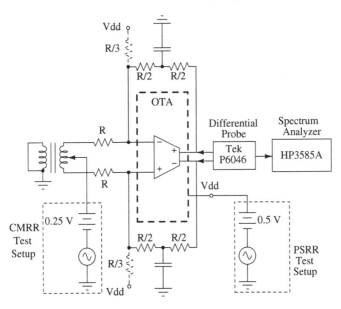

Fig. 2.18: Setup for measuring CMRR and PSRR of the OTAs. The dashed resistor is required for testing the gate-input OTA only.

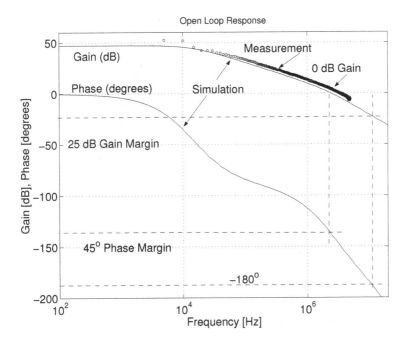

Fig. 2.19: Simulation and measurements of the body-input OTA in open loop.

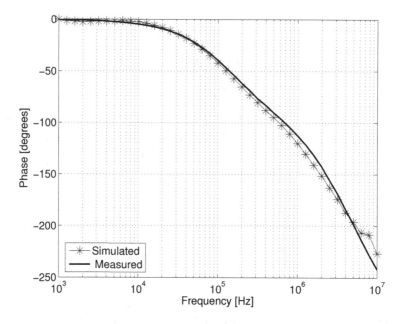

Fig. 2.20: Simulated and measured phase response of the body-input OTA, in open loop. The simulations were adjusted for board parasitic capacitances.

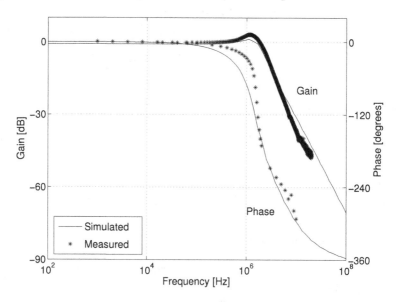

Fig. 2.21: Simulated and measured gain and phase response of the body-input OTA in closed loop.

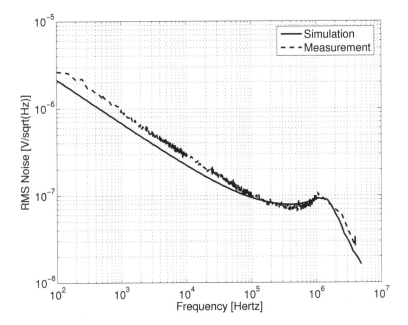

Fig. 2.22: Simulated and measured output noise in the closed loop configuration.

and performance of the gate-input OTA matched well with simulations in the slow process corner.

2.5 Discussion on the two OTA design techniques

The two designed OTAs are compared with measured results of other low voltage CMOS OTA designs in Table 2.4. The gate-input OTA performs better than the body-input OTA in terms of gain-bandwidth and power consumption. This is primarily due to the use of g_m of the input devices as opposed to g_{mb}. The input-referred noise is also smaller in the gate-input OTA. However, when the gate-input OTA is used in feedback, extra resistors are required to set the input common-mode voltage as in Fig. 2.7. This adds resistor white noise directly at the inputs of the OTA and significantly contributes to the total noise of the circuit in feedback. The body-input OTA has a large input common-mode range and has functionality for all input common-mode voltages from 0 to 0.5 V. The common-mode gain increases at low input common-mode voltages. The gate-input OTA, on the other hand, is designed to have large gain only at input common-mode voltages larger than 0.4 V.

Table 2.3: Key parameters of the body-input and gate-input OTAs

Parameter	Body-Input OTA		Gate-Input OTA	
	Measured	Simulated	Measured	Simulated
Nominal supply voltage [V]	0.5	0.5	0.5	0.5
Power dissipation [μW]	110	100	75	100
Area [mm^2]	0.026		0.017	
Offset standard deviation, (20 samples) [mV]	3		2	
Input current @ 27° C [nA]	< 1	0.25	NA	NA
Open-loop DC gain [dB] (diff.)	52	48	62 42*	72 46*
Open-loop unity-gain BW [MHz] (diff.)	2.5	2.4	10.0	15.0
Slew Rate [V/μsec] (diff.)	2.89	2.92	2.0	2.7
Closed-loop unity-gain BW [MHz] (diff.) †	2.2	2.0	5.0	6.5
CMRR ($A_{\text{diff}}/A_{\text{cm}}$) @ 5 kHz [dB]	78	78	74.5	85
CM input to differential output gain [dB]	-63		-56	
PSRR @ 5 kHz [dB]	76	NA	81.4	NA
Input ref. noise @ 10 kHz [nV/$\sqrt{\text{Hz}}$] † (diff.)	280	220	225	120
Input ref. noise @ 1 MHz [nV/$\sqrt{\text{Hz}}$] † (diff.)	80	90	70	100
Load capacitance [pF] (single-ended)	20	20	20	20
Output amp. for 1% HD$_3$ [mV p-p] (diff.) †	400		712	
Output clipping level [mV p-p] (diff.) †	520		752	

* With the gain enhancement biasing disabled.

† Measurements were done in a closed-loop inverting configuration as shown in Fig. 2.17. For the body-input OTA, 10 kΩ resistors were used; for the gate-input OTA, 50 kΩ resistors were used. Additional resistors from the virtual grounds to V_{DD}, of 16.7 kΩ, were used for the gate-input OTA to maintain input and output common-mode voltage levels.

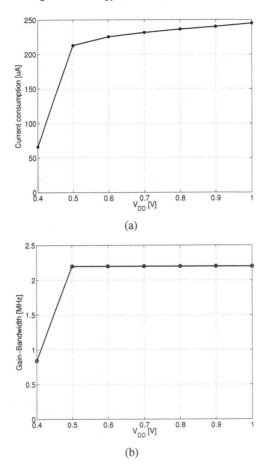

Fig. 2.23: (a) Total current consumption and (b) gain-bandwidth product, for different V_{DD}.

2.6 Design methodology for low V_T devices, without body access

In nano-scale technologies, the V_T of the devices will be much lower than 0.5 V. On the other hand, the use of the bodies of the devices as a back gate might not be feasible because of, either very low body effect [55], or restricted access to the body. In this section, we will briefly explore extensions to the OTA and bias circuit designs presented earlier in this chapter, using low V_T devices, and without requiring access to the bodies of these devices.

2.6.1 Fully differential OTA

The body-input OTA, discussed earlier in Section 2.1, cannot be implemented without access to the body, and hence, will not be discussed in this section.

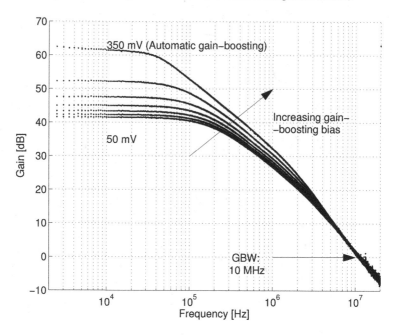

Fig. 2.24: Measured open-loop frequency transfer function of the gate-input OTA for different V_{NR}.

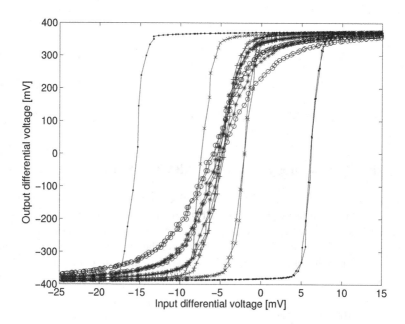

Fig. 2.25: Measured DC input-output transfer characteristics of the gate-input OTA, for different V_{NR}.

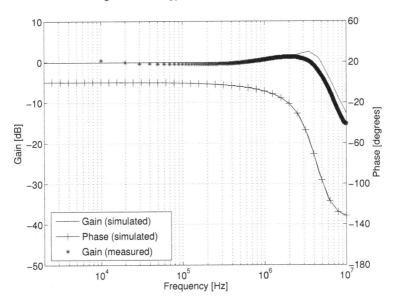

Fig. 2.26: Simulated and measured closed-loop gain and phase response of the gate-input OTA.

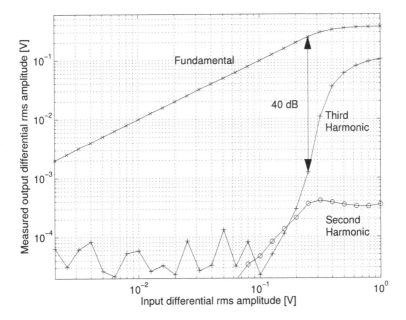

Fig. 2.27: Harmonic distortion in the gate-input OTA, in closed loop, for different input amplitudes.

Table 2.4: Comparison with other low voltage OTA designs

Parameter	[16]	[22]	[23]	[24]	[69]	[37]	Body-input OTA	Gate-input OTA
Supply [V]	1	1	0.8	0.9	1.3	0.9	0.5	0.5
DC Gain [dB]	49	70	53	70	84	59	52	62
GBW [MHz]	1.3	0.2	1.3	6kHz	1.3	4	2.5	10.0
Power [μW]	300	5	–	0.5	460	–	110	75
C_L [pF]	22	7	20	12	–	14	20	20
Single-ended (S) / Diff.(D)	S	S	S	S	S	D	D	D
Technology [μm]	2	0.35	0.5	2.5	0.7	0.5	0.18	0.18
Devices	Lat. BJT	–	Lat. BJT	Depl. MOS	–	–	–	Triple well
100η [1/V]*	9.5	28	–	13	–	–	22.7	133.4

$$^* \; \eta = \frac{\text{GBW} \cdot C_L}{I_{\text{supply}}}$$

A basic input stage of a gate-input OTA, as shown earlier in Fig. 2.8(b), but without the common-mode feed-forward structures, is shown in Fig. 2.28(a). An extension of this OTA stage, that takes full advantage of low-V_T devices, and does not require access to the bodies of the devices, is shown in Fig. 2.28(b). The devices M_{1A} and M_{1B} in Fig. 2.28(a) are each replaced with two devices – M_{1A} is replaced with a combination of M_{11A} and M_{12A}, and M_{1B} is replaced with a combination of M_{11B} and M_{12B}. The earlier small signal analysis of Fig. 2.10 still remains valid with trivial modifications, as shown in Fig. 2.29. The output common-mode voltage of this stage is still controlled by V_{bn}, which now controls the current through M_{12A} and M_{12B}. The devices in this modified input stage are biased in strong inversion, with an overdrive of 0.2 V. This limits the single-ended peak-peak voltage swing at the outputs of the OTA stage to only 100 mV. However, this is satisfactory for the input stage. The output stage of the OTA can be very similar, with the devices biased close to moderate inversion (with V_{GS} of 0.25 V). This enables maximum output swing. The bias voltages at all the nodes in the circuit are indicated. The additional common-mode feed-forward circuitry in Fig. 2.8(b) can be included to improve the common-mode rejection of this OTA input stage.

2.6.2 Bias circuits

The switching threshold voltage of the error amplifiers discussed in Fig. 2.12 are controlled by the voltage V_{amp}. This voltage, applied to the body of the nMOS device in the error amplifier, controls the threshold voltage of the nMOS devices in each error amplifier. Through an active feedback loop, V_{amp} is controlled to set the switching threshold voltage of the error amplifier to $V_{DD}/2$, as shown in Fig. 2.12(a).

Figure 2.30(a) shows a single error amplifier. The nMOS device M_N in Fig. 2.30(a) is replaced with two nMOS devices, M_{N1} and M_{N2}, in Fig. 2.30(b). The input voltage is amplified by the devices M_{N1} and M_P, while V_{amp} controls the current through M_{N2}. Thus V_{amp} controls the switching threshold voltage of the error amplifier in Fig. 2.30(b). This modified error amplifier does not require access to the body node of the nMOS device, and can be used in a feedback loop similar to Fig. 2.12(a) to determine V_{amp}.

With the availability of the error amplifier, biasing networks for V_{bn} and V_L are identical to the circuits described in Sections 2.3.3 and 2.3.2 respectively. The gain-enhancement bias circuit is no longer required in the absence of any such bias voltage for the modified OTA stage in Fig. 2.28(b).

2.7 Summary

In this chapter, design techniques for operational transconductance amplifiers at very low power supplies have been presented. These techniques are not limited by the V_T of the devices. The designs were nominally aimed for a power supply of 0.5 V. However the same techniques could be extended to even lower power supplies at the cost of bandwidth and/or power. Extensions to these techniques have been discussed, which could be used in nano-scale technologies with low-V_T devices without access to the body nodes.

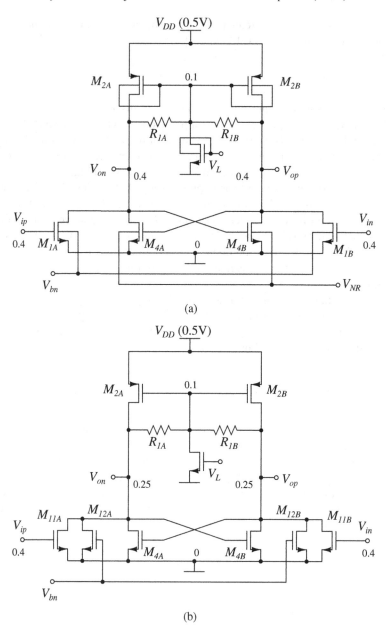

Fig. 2.28: (a) Basic input stage of a gate-input OTA (Fig. 2.8(b)) without the common-mode feed-forward. (b) Basic input stage of a gate-input OTA, using low-V_T devices, without access to the bodies of the devices.

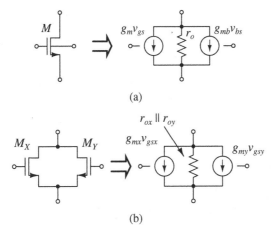

(a)

(b)

Fig. 2.29: Small-signal circuits for (a) device, M, with front and back gates, and (b) modification using two devices, M_X and M_Y, instead of M.

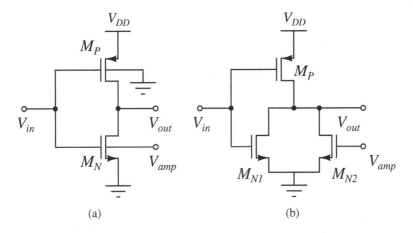

Fig. 2.30: Error amplifier (a) in Section 2.3.1, and (b) modification, without access to the bodies of the devices.

3

Weak Inversion MOS Varactors for Tunable Integrators[1]

A tunable integrator is the basic building block of a tunable filter. Traditionally, tunability can be accomplished using a MOSFET-C structure with MOS devices replacing resistors; or, using switches and banks of resistors and capacitors for discrete tuning; or, using transconductance-C techniques; or, using a varactor-R structure with varactors replacing capacitors. A MOSFET-C structure typically requires the MOS devices to be in strong inversion, which might not be feasible given the ultra-low supply voltage requirement. Using switches in the signal path would require voltage boosting to turn on the switches, which raises reliability concerns. The design of highly linear tunable transconductors at very low supply voltages is very challenging. We have thus investigated the use of varactor-R techniques. Variable capacitors, along with resistors and low voltage OTAs, enable us to build active-RC circuits at 0.5 V. For this, we propose the use of a weak-inversion MOS capacitor as a three-terminal varactor. The capacitance is between the gate and the combination of drain and source, denoted here as C_{gs}, and the tuning voltage is applied at the body, as shown in Fig. 3.1(a). In strong inversion and in accumulation, this capacitance is the oxide capacitance, C_{ox}. In depletion, the intrinsic capacitance is zero as there is no inversion layer. From weak to strong inversion through moderate inversion, the intrinsic C_{gs} changes from zero to C_{ox}. Changing the body voltage changes the device threshold voltage, V_T, and also changes the inversion level of the device. This changes C_{gs} and the device now behaves as a three-terminal varactor.

In this chapter, after a brief theoretical overview in Section 3.1, in Section 3.2, simulation techniques to model the effect of series parasitic resistance are presented and are compared to device measurements. Circuit applications of the varactor are proposed in Section 3.3.

---

[1] ©2005 IEEE. Portions reprinted, with permission, from S. Chatterjee, Y. Tsividis, P. Kinget, "0.5-V Analog Circuit Techniques and Their Application to OTA and Filter Design", *IEEE Journal of Solid State Circuits*, Dec. 2005, vol. 40, no. 12, pp. 2373-2387.

3.1 Brief theoretical overview

The intrinsic gate-source capacitance, C_{gs}, of the device in Fig. 3.1(a), with the drain shorted to the source, is defined as:

$$C_{gs} \equiv -\frac{\partial q_G}{\partial V_S}\bigg|_{V_G,V_B \text{ constant}} \tag{3.1}$$

where q_G is the charge on the gate, and V_S, V_G, V_B are the voltages at the source/drain, gate and body, respectively, with respect to ground.

The MOSFET gate charge can be expressed in a compact form in terms of the surface potential using a charge-sheet approximation [43,70]. In an n-channel device with drain and source connected together, assuming uniform charge along the length of the device, this is given as:

$$q_G = C_{ox}\left(V_{GB} - \psi_S - \phi_{MS}\right) \tag{3.2}$$

where C_{ox} is the oxide capacitance, ψ_S is the channel surface potential, ϕ_{MS} is the work function difference potential. At a given bias voltage, ψ_S can be solved numerically from:

$$V_{GB} = V_{FB} + \psi_S + \gamma\sqrt{\psi_S + \phi_t e^{[\psi_S - (2\phi_F + V_{SB})]/\phi_t}} \tag{3.3}$$

Here, ϕ_t is kT/q, V_{FB} is the flat-band voltage, ϕ_F is the Fermi-potential, $\gamma = \sqrt{2q\epsilon_s N_a}/C'_{ox}$ is the body-effect coefficient, q is the charge of an electron, ϵ_s is the permittivity of silicon, N_a is the acceptor doping concentration in the channel and C'_{ox} is the oxide capacitance per unit area. C_{gs} can be derived from (3.1) and (3.2), (3.3).

A long-channel nMOS device (15 fingers, each of width 100 μm, length 20 μm) was fabricated in a 0.18 μm triple-well CMOS process, with the body accessible through the p-well and the drain and source shorted together. The normalized gate-source capacitance of this device, for different bias voltages, is shown in Fig. 3.1(b) and 3.1(c). For an operating point for V_{GS} of about 0.25 V, over the range of V_{GB} from -0.1 V to 0.4 V, the capacitance varies by a factor of 4. At an operating point for V_{GS} of 0.15 V, the capacitance varies from 0 to $0.1 C_{ox}$. For maximum tuning range, the MOS device can be used at a V_{GS} of 0.15 V in this fabrication process.

3.2 Device measurements and modeling

The varactor was measured and characterized using the Agilent 4284A LCR-meter. Fig. 3.2 depicts the measurement setup. The measured effective capacitance, C_{eff} and series resistance, R_{eff} are defined in Fig. 3.3(a). The effective capacitance is shown in Fig. 3.3(b) as a function of V_{GS} for different V_{GB}. Due to the lack of a strong inversion layer, there is significant resistance in series with the capacitance

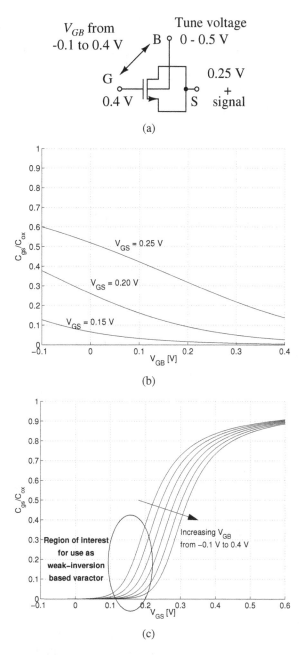

Fig. 3.1: (a) Gate (G) to Source/Drain (S) capacitance tuned by Body (B). (b) Normalized capacitance as a function of V_{GB} for different V_{GS}. (c) Normalized capacitance as a function of V_{GS} for different V_{GB}. $N_a = 3.5 \times 10^{17} \text{cm}^{-3}$, $V_{FB} = -1$ V.

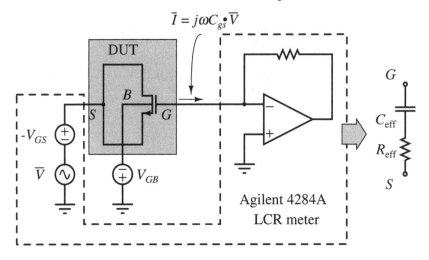

$$\bar{I} = j\omega C_{gs} \cdot \bar{V}$$

Fig. 3.2: C_{gs} measurement setup using Agilent 4284A LCR meter.

which reduces the quality factor of the capacitor. The resistance measured in series with the capacitance at 1 MHz is shown in Fig. 3.3(c).

The significant series resistance can be attributed to distributed effects – the source and the drain are connected outside the device, but internally there is a large channel resistance between them. A distributed model used to predict this resistance is shown in Fig. 3.4. Techniques to accurately predict this series resistance have been discussed in [71]. More details on the varactor modeling are given in Appendix A.

3.2.1 Closed-form model

We define the gate-source transadmittance, $\overline{Y_{gs}}$, by $\overline{I_g}/\overline{V_s}$, where $\overline{V_s}$ is a voltage phasor applied to the source/drain terminal and $\overline{I_g}$ is a current phasor leaving the gate terminal, with V_B and V_G constant. Distributed analysis techniques, as discussed in [71], show that the effective transadmittance, $\overline{Y_{gs}(j\omega)}$, (not including overlap capacitance), starting from the model in Fig. 3.4 and letting the number of sections go to infinity, is given by:

$$\overline{Y_{gs}(j\omega)} = j\omega C_{gs} \frac{\tanh \alpha/2}{\alpha/2} \qquad (3.4)$$

where $\alpha = \sqrt{j\omega R_{ds}(C_{gs} + C_{bs})}$. C_{gs}, C_{bs} and R_{ds} are the lumped gate to source/drain capacitance, body to source/drain capacitance, and source to drain resistance respectively. A detailed derivation of (3.4) has been included in Appendix A. The intrinsic transadmittance, $\overline{Y_{gs}}$, can now be combined with the overlap capacitance, and then transformed to extract the effective series resistance and series capacitance. The effective gate-source capacitance is given by:

$$C_{\text{eff}} = -\frac{1}{\omega \cdot \text{Im}\{1/[Y_{gs}(j\omega) + 2j\omega C_{\text{overlap}}]\}} \qquad (3.5)$$

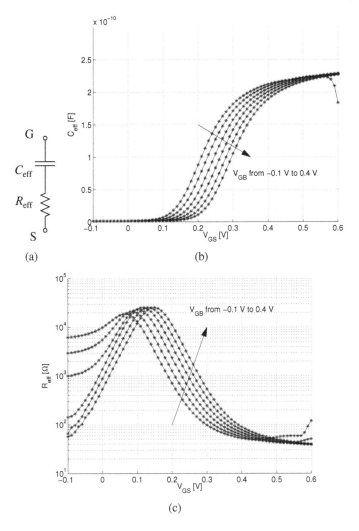

Fig. 3.3: (a) Lumped model of the capacitor showing effective capacitance and series resistance. (b) Measured effective capacitance, C_{eff}, and (c) series resistance, R_{eff}, as a function of V_{GS} for different V_{GB}. A measurement frequency of 1 MHz was used.

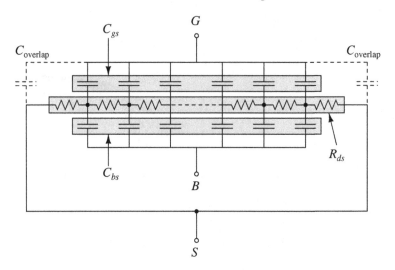

Fig. 3.4: Conceptual model of the capacitor. C_{gs}, C_{bs}, R_{ds} are the lumped gate to source/drain capacitance, the lumped body to source/drain capacitance, and the lumped drain to source resistance.

The effective series resistance is given by:

$$R_{\text{eff}} = \text{Re}\left\{ \frac{1}{Y_{gs}(j\omega) + j\omega 2 C_{\text{overlap}}} \right\} \tag{3.6}$$

If R_{ds}, C_{bs}, C_{gs} at any bias point are found using a given MOS model, then (3.4), through (3.5) and (3.6) can be used to extract the effective series resistance and capacitance at the frequency of interest. Fig. 3.5 shows a flow-chart that can be used to compute the effective series resistance and capacitance at the frequency of interest.

3.2.2 Channel segmentation

The distributed effect can also be quickly simulated using a standard circuit simulator by using channel segmentation [43, 72], as in Fig. 3.4. A long-channel MOS device can be viewed as a series connection of several short intrinsic MOS devices with the drains and sources of adjacent devices connected to each other. If a device of length L μm is broken into N segments, each of the channel segments has length of L/N μm. As the number of segments, N, becomes larger, this will approach a truly distributed model. It has been shown in Appendix A, that for a segmented model, the number of segments, N, should be such that at the frequency of interest, $\omega R_{ds}(C_{gs} + C_{bs})/2N^2 \ll 1$. As a matter of practical importance while using this technique for simulation, spurious overlap capacitances and junction diodes need to be turned off for the sub-devices that are not connected to the drain or source of the composite device, as shown in Fig. 3.6. For the BSIM3 model [73], nulling the parameters

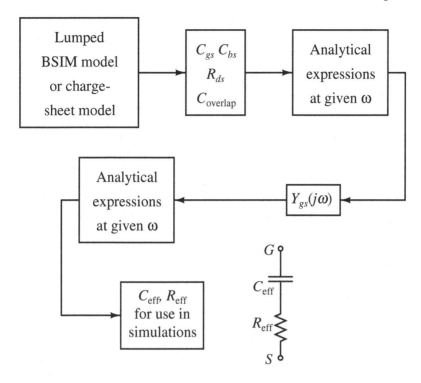

Fig. 3.5: Flow-chart depicting the procedure to compute C_{eff} and R_{eff}.

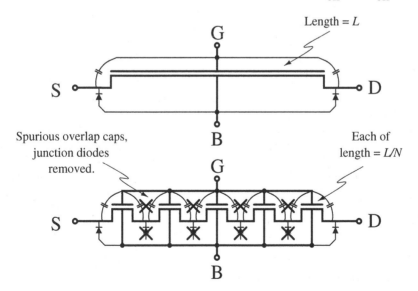

Fig. 3.6: Simulation technique using channel segmentation.

CGS0, CGS1, CGD0, CGD1 and CGB0 will zero the overlap capacitance. Setting CJ, CJSW, CJSWG, TCJ, TCJSW, TCJSWG, to zero, nulls the source/drain to body diode junction capacitance. Setting JS, JSSW, to zero ensures that the DC current through the junction diode is also zero, so that the diode is disabled.

3.2.3 Comparison between measured results and simulations

Figure 3.7 shows the measured effective capacitance in comparison with modeled results. A charge-sheet model, in combination with the distributed technique discussed in Section 3.2.1 closely approximates the measured data. Measured data, as well as modeled results, are invariant over the range of frequencies from 1 kHz to 1 MHz. Measurements were limited by the frequency range of the Agilent 4284 LCR-meter. A segmented simulation approach does not track the measured capacitance well, because of the limitations of the BSIM model used to simulate each of the individual segments.

Figure 3.8 shows the measured R_{eff} in comparison with modeled results. The series resistance increases almost exponentially as V_{GS} decreases. When the intrinsic C_{gs} is very small, the admittance offered by the small overlap capacitance becomes comparable to the admittance of the series resistance and the intrinsic C_{gs}. As a result, the observed series resistance, R_{eff}, starts to drop. A BSIM model, in combination with the distributed technique closely approximates the measured data. The charge-sheet model did not track the measured series resistance, and is not shown, because of limitations in modeling electron mobility in the channel. R_{eff} predicted by a segmented simulation also closely approximates the measured data.

3.3 Circuit applications

3.3.1 Discrete prototype using the varactor

The varactor can be used in any low frequency application with the appropriate bias voltages applied. The individual device, in the presence of parasitics, was used to build a tunable 25-30 kHz notch filter along with off-the-shelf discrete components. The bias-voltage setup, as well as the complete prototype circuit schematic is shown in Fig. 3.9. The attained depth of the notch was a direct result of the effective resistance in series with the capacitance, as well as the opamp gain, and has been verified with simulation using a segmented varactor model. The measured frequency response of the notch filter for different V_{GB} is shown in Fig. 3.10 and is compared against simulations in Table 3.1. Excellent agreement is obtained between the simulated and measured results.

3.3.2 Application of the varactor in an integrated setting

The implementation of a low voltage tunable damped integrator, using the proposed varactors and a gate-input OTA, is shown in Fig. 3.11. Resistors to V_{DD} at the inputs

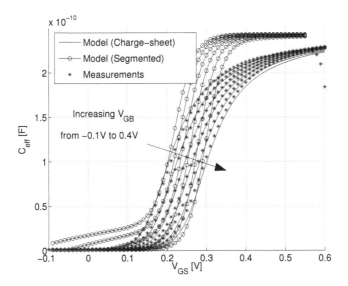

Fig. 3.7: Comparison of measured C_{eff} to the C_{eff} predicted by (–) a combination of the distributed model and the charge-sheet model (see Fig. 3.5), and also to the C_{eff} predicted through (-o-) segmentation (see Fig. 3.6), for a device of width 1.5 mm, length 20 μm, in a 0.18 μm triple-well CMOS process. Over the measured frequencies of 1 kHz to 1 MHz, variations in capacitance are very small.

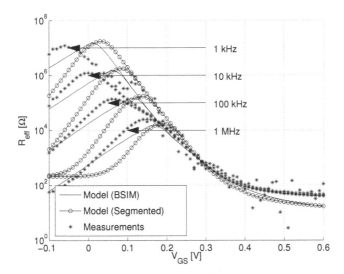

Fig. 3.8: Comparison of measured R_{eff} to the R_{eff} predicted by a (–) combination of the distributed model and the BSIM model (see Fig. 3.5), and also to the R_{eff} predicted through (-o-) segmentation (see Fig. 3.6), for V_{GB} of 0.4 V, at different frequencies.

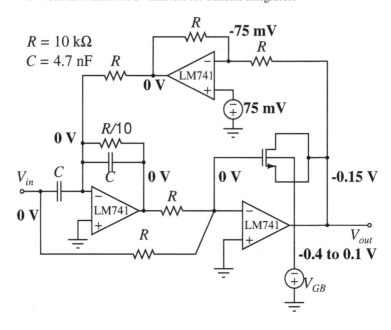

Fig. 3.9: Schematic of a notch filter built with off-the-shelf components and 0.18 μm varactor test devices.

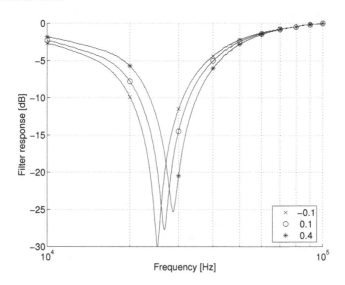

Fig. 3.10: Frequency response of the notch filter at different V_{GB} tuning voltages.

Table 3.1: Simulations and measurements for the notch filter

V_{GB}	Notch freq. [kHz]		Notch depth [dB]	
[V]	(sim.)	(measd.)	(sim.)	(measd.)
-0.1	24.1	25.1	-29.7	-30.1
0.1	27.2	26.8	-27.0	-27.6
0.4	29.0	28.8	-25.3	-25.2

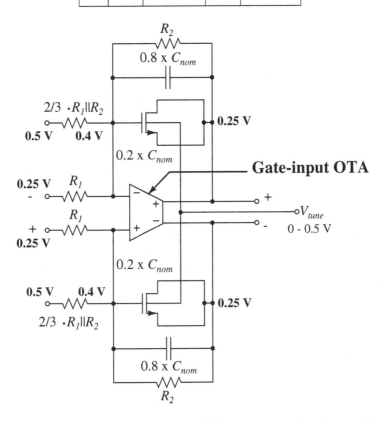

Fig. 3.11: 0.5 V tunable damped integrator. DC common-mode voltages of the different nodes are shown in bold.

of the gate-input OTA (see Fig. 2.7 and relevant discussion in Section 2.2) maintain a common-mode level of 0.4 V at the virtual grounds when the integrator input and output common-mode voltages are 0.25 V. These resistors do not affect the transfer function of the integrator.

Note that for a high-gain OTA, the gate of the varactor is at a virtual ground, and the body is at a DC voltage. The current leaving the gate of the varactor is given by $-\frac{dq_G}{dt}$. From (3.1),

$$i_G = C_{gs} \frac{dV_S}{dt}$$

or, in the frequency domain,

$$\overline{I_g(j\omega)} = j\omega C_{gs} \overline{V_s(j\omega)}$$

As a result, the voltage at the output of the integrator depends only on C_{gs}, and is not affected by $C_{sg}, C_{gb}, C_{bg}, C_{sb}$ and C_{bs}. Note that the above expression is valid only when the OTA has high gain, and the output voltage is a small signal, i.e, voltage swing at the output (or source/drain) does not change C_{gs} appreciably.

With a V_{GS} of 0.15 V, the proposed varactor has low capacitance density but has a very large tuning range. In the 0.18 μm CMOS process its density is about 0.3 fF/μm^2. This limitation is offset by adding a fixed capacitor in shunt with the variable capacitor. Any available fixed capacitor type can be used and to satisfy the area restrictions for our prototype, native devices (zero-V_T) were used as fixed capacitors; they operate in strong inversion under the above biasing conditions and have a density of 8 fF/μm^2. This gives us an overall capacitance density of 1.3 fF/μm^2. An added benefit is a better quality factor for the composite capacitance. If the fixed capacitor is linear, the linearity of the composite varactor is improved. In this implementation, the fixed capacitor was chosen to be 80% of the nominal capacitance required, and the varactor was sized to be 20% of the nominal capacitance at the center of its range. This enables a tuning range of $\pm 20\%$. This tunable integrator prototype was used to build a tunable 5th order low-pass elliptic filter, to be discussed in Chapter 4.

3.4 Summary

In this chapter, a weak inversion MOS varactor has been demonstrated and it has been applied in ultra-low voltage analog filter circuits down to 0.5 V. The varactor has been modeled and distributed effects have been analyzed to derive an expression for the parasitic series resistance. Device characterization results correspond well to the presented models. A segmented device approach has been used to simulate the distributed nature of the varactor in practical circuit designs. Circuit examples have been presented and correspondence between measurements and simulation results have been obtained.

4

A 0.5 V 5th-Order Low-Pass Elliptic Filter[1]

4.1 Filter topology

To demonstrate the capabilities and synergy of the proposed ultra-low voltage design techniques in Chapters 2 and 3, we designed a 5th-order low-pass elliptic filter with a 135 kHz cut-off frequency. For minimum sensitivity requirements, a leap-frog topology was used. The design in [74] was frequency-scaled to 135 kHz, such that the signal amplitude maxima at all OTA outputs are at the same level. The filter characteristics has a pass-band ripple of 0.1 dB, a stop-band rejection of at least 35 dB, and two zeros in the stop-band – at 180 kHz and at 280 kHz. To obtain an accurate transfer characteristic, the OTA should have substantial open-loop gain all the way to 280 kHz, the second zero of the filter. The proposed amplifier has a worst-case gain of 20 dB at 280 kHz, which is sufficient. The filter resistors and capacitors were scaled so that the total noise contributions from the OTAs and from the resistors, integrated in the pass-band, are equal. The schematic is presented in Fig. 4.1, with component values in Table 4.1. The sizing of the individual gate-input OTAs is scaled depending on the loading requirements, by connecting multiple units in parallel, with all internal nodes of the OTAs connected to each other. This allows comparable phase and distortion performance for all five stages. As a result, the first integrator stage of the filter uses two OTA units in parallel, the second uses two OTA units, the third uses four OTA units, the fourth uses one OTA unit, and the fifth uses two OTA units in parallel. Fig. 4.2 shows the ideal filter characteristics of the LC-prototype and the simulated characteristics of the low voltage design.

The capacitors are replaced with weak inversion low voltage MOS varactors (see Chapter 3). The simulated overall dynamic range – ratio of input r.m.s value at which there is 1% THD to the input referred noise – is 57 dB. The varactors contribute substantially to the distortion of the circuit. Using ideal linear capacitors instead of varactors, the simulated dynamic range is 69 dB, whereas, using ideal OTAs instead of the low voltage OTAs, the simulated dynamic range is only 58 dB.

Table 4.1: Filter resistor and capacitor values

Capacitor	Value	Resistor	Value
C_1	9.2 pF	R	300 kΩ
C_2	13.1 pF	R_1	30 kΩ
C_3	20.2 pF	R_2	50 kΩ
C_4	8.8 pF	R_3	40 kΩ
C_5	5.3 pF	R_4	66.7 kΩ
C_6	1.8 pF	R_5	57.2 kΩ
C_7	5.7 pF		
C_8	2.9 pF		

4.2 On-chip PLL-based automatic frequency tuning loop

An ultra-low voltage, voltage-controlled oscillator, shown in Fig. 4.3, was built using tunable integrators with the resistors and capacitors matched to those in the filter. The oscillator frequency, f_0, is chosen to be close to the second zero of the filter, 280 kHz. A three-stage oscillator was chosen in preference to a double-integrator oscillator. The OTAs have enough gain-bandwidth to set a phase lag of 60° per stage along with the required gain of greater than 1, at f_0, to reliably sustain oscillations. The oscillator has a nominal frequency of oscillation given by:

$$f_0 = \frac{\sqrt{3}}{2\pi R_a C_a}$$

and oscillations are possible only when $R_a \geq 2R_b$. In Fig. 4.3, R_a is 427 kΩ, R_b is 207 kΩ, R_{VDD} is 93 kΩ, and C_a is 2.3 pF.

A phase-locked loop is built around the VCO using an XOR gate as a phase detector. A pass-transistor XOR gate is used and its schematic is shown in Fig. 4.4. A pass-transistor XOR gate does not have any stacking of devices, is intrinsically fast, and is suitable for this application. With a 0.5 V power supply voltage, the devices in the XOR gate are in weak inversion, but they are still sufficiently fast for 280 kHz. This detector compares the VCO frequency to an external reference clock and controls the body voltage of the capacitors in the VCO, till the PLL is in lock. The same capacitor body voltage is applied to the filter as well. The filter varactors and resistors are matched to those in the VCO, and in this manner the corner frequency of the filter is tuned. For characterization flexibility, a first-order PLL loop filter is built externally using discrete components.

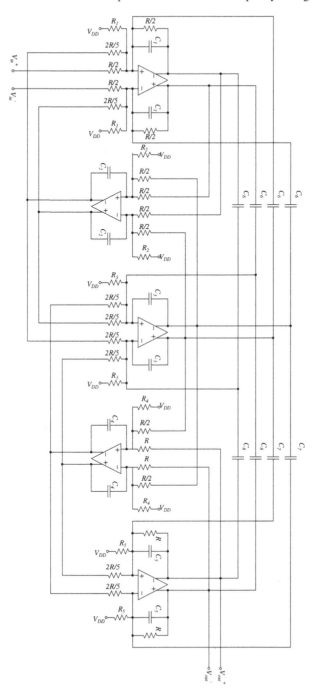

Fig. 4.1: Low voltage 5th-order low-pass elliptic filter.

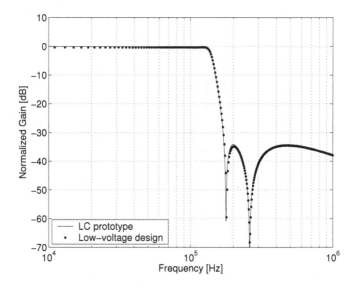

Fig. 4.2: Simulated filter characteristics of (–) LC prototype, and (·) the low voltage design.

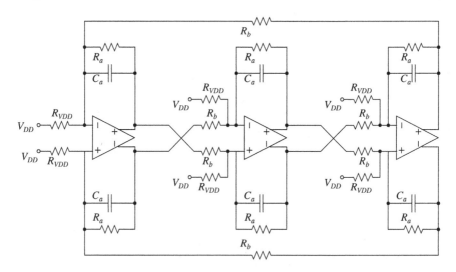

Fig. 4.3: Three stage low voltage oscillator schematic.

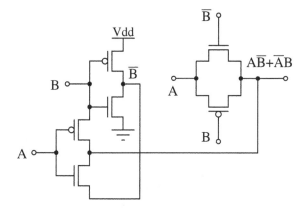

Fig. 4.4: Pass-transistor fast XOR gate.

4.3 Layout and prototype chip

The full chip, shown in Fig. 4.5, containing the 5th-order filter, VCO, bias circuits and phase detector was laid-out and fabricated in a 0.18 μm CMOS process, taking advantage of triple-well nMOS devices, high-resistivity resistors, and MIM capacitors. A total of eleven unit-OTAs were used for the 5th order low-pass filter, and three unit-OTAs were used for the three-stage VCO.

The filter capacitors (varactors) were broken down into unit capacitances of 1.4 pF (nominal) each. The remainder of the capacitance required to complete the implementation of each of the filter capacitors was designed such that the area to perimeter ratio was identical to that of the unit capacitor. For example, to design C_3 (see Table 4.1) of 20.2 pF, 14 unit capacitors of 1.4 pF each were added in shunt with a capacitor of 0.6 pF. The 0.6 pF capacitor was designed to have an area to perimeter ratio equal to that of the 1.4 pF unit capacitor. Careful design of each of the capacitors in the filter as described above, improved the chances that the filter characteristics would be as close to the ideal as possible [47, 75].

As discussed earlier in Section 3.3.2, each of the varactors was implemented using a shunt combination of zero-V_T devices and standard MOS capacitors. The zero-V_T devices were used as fixed capacitors, whereas, the standard MOS devices were used as variable capacitors, with tuning achieved through the body.

The total active chip area was 1 mm $\times$ 1 mm. The chip micro-graph is shown in Fig. 4.6. Only one external voltage reference of 0.4 V is used for biasing. An external frequency reference of 280 kHz is used to tune the filter. To account for mismatches between the OTAs in the filter and the replica OTAs in the biasing circuits, the bias voltage V_{NR} (see Section 2.3.4) was scaled down using resistors, and applied to the filter OTAs.

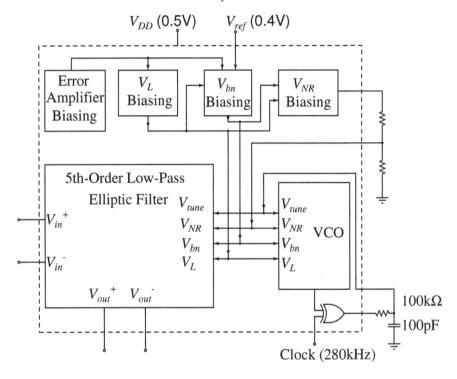

Fig. 4.5: Block diagram of the full chip.

4.4 Characterization Results

4.4.1 Test set-up

A center-tap transformer was used to convert a single-ended signal to a differential signal with a common-mode voltage of 0.25 V. A Tektronix P6046 differential probe was used to measure the differential outputs of the filter. Extensive measurements were taken using the HP3585A Spectrum Analyzer. Results are reported in several following figures and in Table 4.2.

4.4.2 Frequency response

The measured filter frequency response agrees closely to simulation as shown in Fig. 4.7(a). The pass-band ripple is 0.1 dB. Fig. 4.7(b) shows the frequency response of the filter with and without automatic gain-enhancement. With gain-enhancement, the filter response improved by 1 dB in the pass-band. With a 10 kHz common-mode tone, the measured CMRR was 65 dB. With a 10 kHz tone on the power supply, the measured PSRR was 43 dB. Better PSRR can be achieved if a separate regulated power supply is used for all the biasing resistors, R_b, as in Fig. 2.7.

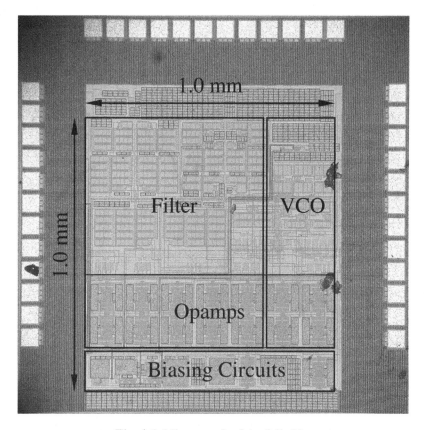

Fig. 4.6: Micro-graph of the full chip.

4.4.3 Noise

The measured output noise is compared to simulation in Fig. 4.9. At low frequencies, the output noise is dominated by $1/f$ noise; the small flat region in the pass-band of the filter is a result of white noise. The peaking at around 135 kHz is a result of the filter topology and transfer function. The observed $1/f$ noise corner is about 40 kHz. The OTAs contribute to the noise at lower frequencies while the filter resistors contribute dominantly beyond the $1/f$ noise corner frequency.

4.4.4 Distortion and characterization over tuning range

Harmonic distortion was measured for an in-band tone of 20 kHz for which the harmonics are also in-band, and for an in-band tone of 100 kHz for which the harmonics are out-of-band. In the worst case, a 1% THD was observed at an input r.m.s differential voltage of 50 mV. Fig. 4.8(a) shows the second and third-order harmonic distortion components for an in-band input tone of 20 kHz.

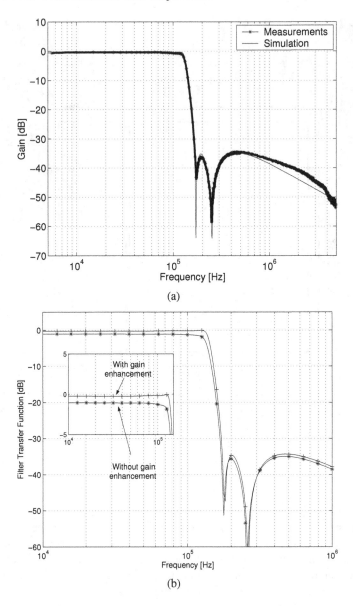

(a)

(b)

Fig. 4.7: Frequency response of the filter. (a) Simulation and measurement. (b) With and without gain enhancement.

Intermodulation measurements were taken for a pair of in-band input tones at 20 kHz and 25 kHz such that their intermodulation products are at 15 kHz and 30 kHz, and also for a pair of out-of-band input tones at the two peaks in the stopband, nominally at 180 kHz and at 460 kHz. The respective input r.m.s differential IIP3s observed are -3 dBV and 5 dBV. Fig. 4.8(b) shows the third order intermodulation products at 15 kHz and 30 kHz for two input tones at 20 kHz and 25 kHz.

For this discussion, dynamic range is defined as the ratio of the input differential r.m.s voltage at which there is 1% THD (worst case) to the input integrated noise [74] from 1 kHz to 150 kHz. The observed dynamic range was 56.6 dB, which is in close agreement with the simulated dynamic range of 57 dB.

To observe the contribution of the varactor to the distortion of the filter, the PLL was deactivated, and the distortion was measured for different tuning voltages applied through the body of the varactors. For a tuning voltage of 0 V, the capacitance non-linearity is reduced as the capacitors consist of fixed strong inversion transistors in shunt with only the overlap capacitance of the varactors. The observed dynamic range is 61 dB at this tuning voltage. For a tuning voltage of 0.5 V, the observed dynamic range decreases to 55 dB.

The frequency response of the filter at different tuning voltages, with the PLL deactivated, is shown in Fig. 4.11. The performance of the filter was evaluated at different tuning voltages and is summarized in Table 4.3. The depths of the filter notches in Fig. 4.11 depend on the OTA gain and on the quality factor of the capacitors. For the 5th-order filter, to obtain the effect of the series resistance, the varactor was simulated by including a calculated resistance (see Fig. 3.5) in series with the device, as well as by using a segmented model (see Fig. 3.6). The simulated depths of the first notch in Fig. 4.11, in either case, at body tuning voltages of 0 and 0.5 V are -53 and -46 dB, as opposed to measured -50 and -43 dB, respectively. The difference of 3 dB in both cases can be attributed to differences in the OTA gain-bandwidth product.

4.4.5 Performance at different power supply voltages

The performance of the filter was evaluated at different power supply voltages and is summarized in Table 4.2. The PLL locks under nominal conditions at these different power supply voltages and all measurements, except the filter tuning range, were taken with the PLL active. The filter transfer function for different power supply voltages is shown in Fig. 4.10(a).

4.4.6 Performance over different chips

The nominal performance was evaluated for a batch of 20 chips. The filter -3 dB cut-off frequency had a standard deviation (σ) of 1.3%. The measured filter characteristics of different chips are shown in Fig. 4.10(b) and are found to agree closely. The mean current consumption at 0.5 V was 2.2 mA with a standard deviation of 0.1 mA.

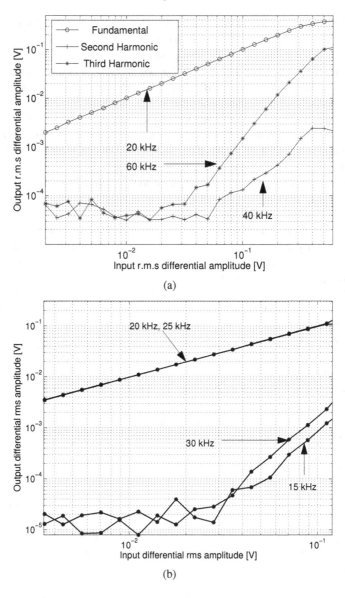

(a)

(b)

Fig. 4.8: (a) Harmonic distortion for an in-band input tone at 20 kHz, for a 0.5 V power supply. (b) Third-order intermodulation for two input tones at 20 kHz and 25 kHz, for a 0.5 V power supply.

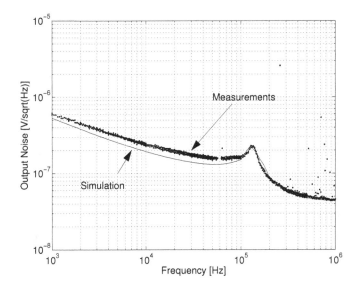

Fig. 4.9: Simulation and measurement of filter noise for 0.5 V supply voltage (Output noise).

4.4.7 Performance over temperature

A Delta Design 3900C temperature chamber was used to characterize the filter at different temperatures. The chip was fully functional from 5° to 85° C and no leakage problems were observed. The characteristics of the filter at different temperatures is shown in Fig. 4.12. The -3 dB cut-off frequency had a variation of $\pm 8\%$ over this range. This somewhat large deviation was diagnosed to be a systematic shift in the VCO characteristic over temperature, caused by relative amplitude variations, which are significant at the 0.5 V supply voltage used. This can be fixed with additional oscillator amplitude stabilization circuitry. At 100° C, the VCO ceased to oscillate, and the PLL was out of lock. However, the filter was still fully functional. Additional oscillator amplitude control circuitry will be necessary to fix this problem. The nominal current consumption at different temperatures is shown in Table 4.4. Device threshold voltages decrease with increasing temperature. The biasing circuits establish fixed DC bias voltages independent of temperature. As a result, current consumption increases with temperature.

4.5 Summary

In this chapter, circuit techniques have been introduced to design a fully differential 5th-order active-RC elliptic low-pass filter, that can operate at a 0.5 V power supply. A low voltage VCO is designed to match the filter, and a PLL with the VCO embedded is used to tune the frequency response of the filter. The design is true low voltage

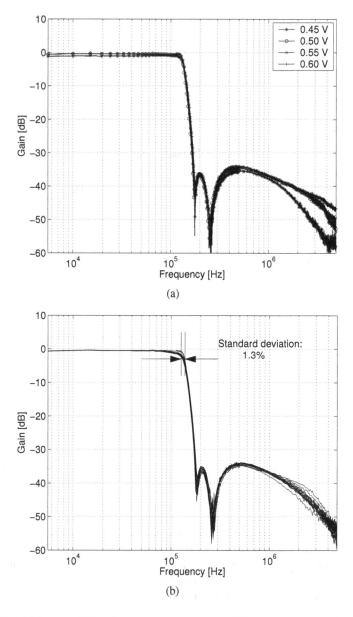

Fig. 4.10: (a) Measured filter frequency response at different power supply voltages (PLL active). (b) Measured frequency response for 20 chips (PLL active).

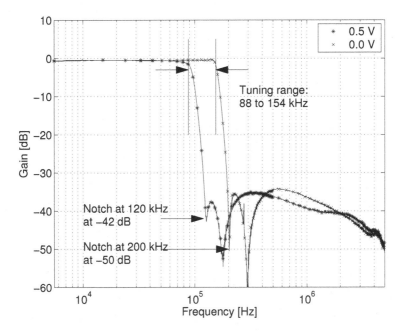

Fig. 4.11: Measured frequency response at different tune voltages (PLL disabled).

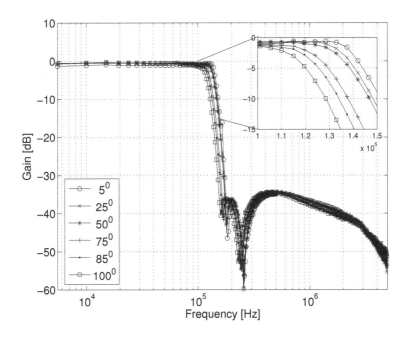

Fig. 4.12: Measured frequency response of the filter at different temperatures from 5°C to 100°C.

Table 4.2: Performance at different power supply voltages, with the PLL enabled

V_{DD} [V]	0.45	0.5	0.55	0.6
-3 dB cut-off frequency [kHz]	135.0	135.0	135.0	135.0
Total current [mA]	1.5	2.2	3.3	4.3
Noise[†] [μV rms]	87	74	68	65
Input[‡] [mV rms] (20kHz / 1% THD)	80	80	80	80
Input[‡] [mV rms] (100kHz / 1% THD)	50	50	50	50
In-band IIP$_3$ [dBV]	-5	-3	-3	-3
Out-of-band IIP$_3$ [dBV]	3	5	3	5
Dynamic range[*] [dB]	55.2	56.6	57.3	57.7
Tuning range [kHz] $V_{tune} = V_{DD}$	96	88	84	69
$V_{tune} = 0.0$ V	153	154	148	150
PLL tone feed-through [μV rms] @280 kHz	104	85	72	72

[†] : Input-referred, differential, integrated over 1kHz to 150kHz
[‡] : Differential
[*] : Ratio of input @100kHz for 1% THD to integrated input noise

– all nodes in the circuit operate within the power rails at all times. The PLL-tuned filter has a 135 kHz cut-off frequency, has 57 dB of dynamic range, and consumes 2.2 mA from a 0.5 V power supply.

Table 4.3: Performance of the filter with the PLL disabled, at different tuning voltages

Tuning voltage [V]	0.0	0.3	0.5
Power supply voltage [V]	0.5	0.5	0.5
Total current [mA]	2.2	2.2	2.2
Cut-off frequency [kHz]	94.0	135.0	158.5
Noise[†] [μV rms]	75	74	70
Input[‡] [mV rms] (20 kHz / 1% THD)	159	80	50
Input[‡] [mV rms] (100 kHz / 1% THD)	80	50	40
In-band IIP_3 [dBV]	1	-3	-1
Out-of-band IIP_3 [dBV]	7	5	3
Dynamic Range* [dB]	60.6	56.6	55.1

[†] : Input-referred, differential, integrated over 1 kHz to 150 kHz
[‡] : differential
* : Ratio of input @100kHz for 1% THD to integrated input noise

Table 4.4: Measured total current consumption at different temperatures

Temperature [^{o} C]	5	25	50	85
Total current [mA] (measured)	1.8	2.2	2.5	3.0
Total current [mA] (simulated)	1.9	2.2	2.6	2.9

5

A 0.5 V Track-and-Hold (T/H) Circuit[1]

5.1 Introduction

The ability to track and then sample an analog signal is an essential function in many signal acquisition interfaces. In such interfaces, analog waveforms need to be sampled and held to within the accuracy of the system, prior to quantization by either sequential or parallel means. In pipelined A/D converters, e.g. in [76], the input signal sample has to be held constant over the duration of each step in the A/D operation. As such, track-and-hold (T/H) circuits are regularly used in an analog-to-digital (A/D) converter.

In this chapter, a brief overview of the basic T/H architectures and design in the context of operation at an ultra-low power supply voltage are presented in Section 5.2. A fully differential T/H circuit for 0.5 V operation is proposed in Section 5.3, followed by design details and measurement results in Section 5.4. Concluding remarks are presented in Section 5.5.

5.2 T/H operation at ultra-low voltages

Figure 5.1 shows a very simple track/hold circuit. The input waveform is tracked at the output when the switch is ON. When the switch turns OFF, the input voltage is held on the capacitor C_S and appears at the output. As shown in Fig. 5.1, the switch can be implemented using a single nMOS device, or a single pMOS device, or a transmission gate, which is a parallel combination of an nMOS and a pMOS device.

In Fig. 5.2, the conductance of a transmission gate switch, G, is plotted at different switch input voltages. V_{Tn} and V_{Tp} are the threshold voltages of the nMOS and pMOS devices respectively. In Fig. 5.2(a), where the power-supply voltage is more than the sum of V_{Tn} and $|V_{Tp}|$, the conductance of the switch is more than

---

[1] ©2007 IEEE. Portions reprinted, with permission, from S. Chatterjee, P. Kinget, "A 0.5-V 1-Msps Track-and-Hold Circuit with 60-dB SNR", *IEEE Journal of Solid State Circuits*, Apr. 2007, vol 42, no. 4, pp. 722-729.

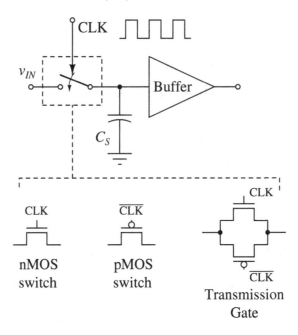

Fig. 5.1: Basic track and hold circuit, with input and output waveforms.

the minimum conductance, G_{min} through the entire range of v_{IN}. Therefore, the allowed input swing is from 0 to V_{DD}. However, in low-voltage circuits where the power-supply voltage is less than the sum of V_{Tn} and $|V_{Tp}|$, e.g. in Fig. 5.2(b), the conductance of the switch is not greater than G_{min} throughout the range of v_{IN}. The nMOS device is ON when v_{IN} is close to 0 (indicated as nMOS swing), the pMOS device is ON when v_{IN} is close to V_{DD} (indicated as pMOS swing). Neither device is ON when v_{IN} is in the middle of its range. As a result, when $V_{DD} < V_{Tn} + |V_{Tp}|$, the T/H circuit of Fig. 5.1 will not be available for use.

For a power supply voltage of 0.5 V, and $V_{Tn}, |V_{Tp}| \approx 0.2$ V a transmission gate switch could possibly be used. However this is difficult too, because the source of the switching transistor can be at a very different voltage from the substrate. The device threshold voltages can vary as much as 100 mV over the possible signal range [5] for typical process parameters, and in effect, the T/H circuit of Fig. 5.1 will not be available for use, even with low device threshold voltages.

The well known approach to overcome this problem is to use internal voltage boosting [27–33]. In some cases the clock voltage is doubled, and that can lead to reliability issues. Other boosting techniques maintain a constant V_{GS}, but could impose a high-voltage glitch. Switched-opamp circuits [34] are particularly attractive and well suited for low power supply voltages. Fully-differential switched-opamp circuits can be used to implement discrete-time transfer functions, and can be integrated in $\Sigma\Delta$ A/D converters. However, T/H and S/H operation cannot be imple-

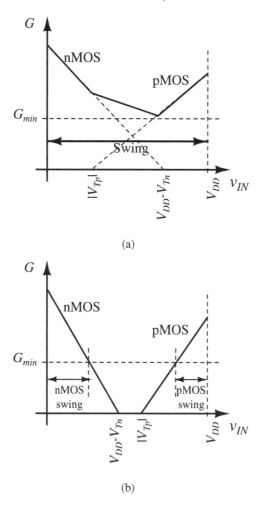

Fig. 5.2: Approximate conductance of a transmission-gate switch, G, as a function of input voltage. (a) Sufficient voltage headroom. (b) No headroom.

mented using the switched-opamp technique. Circuits such as pipelined A/D converters require S/H operation at the input.

5.3 Fully-differential 0.5 V T/H circuit

One way to avoid the problem of little or no voltage headroom for the switches, is to configure circuits in such a way that the switches do not see full swing signals. Placing them at the virtual ground associated with the input of an opamp in feedback configuration is a possibility, e.g in [77, 78]. The single-ended T/H circuit in [77], shown in Fig. 5.3, presents opportunity for use with the OTAs presented in [79]. ϕ_1 and ϕ_2 are non-overlapping clock waveforms. Assuming R_1 and R_2 are equal, when ϕ_1 is high, the feedback around the OTA through the resistors is closed, and v_{out} tracks $-v_{in}$. During this track phase, assuming the switch S_1 presents a very small impedance, the frequency domain transfer function from v_{in} to v_{out} is given by:

$$\frac{V_{out}(s)}{V_{in}(s)} = -\frac{R_1}{R_2} \cdot \frac{1 + sC_2R_2}{1 + sC_1R_1} \tag{5.1}$$

If R_1C_1 is equal to R_2C_2, the pole and the zero cancel out. The use of C_2 will thus, allow the circuit to have a fast response and greatly improve the speed of the T/H circuit. When ϕ_2 is high, the resistive feedback is opened and the capacitor, C_1, across the OTA, holds the voltage at v_{out}. While S1 is open, the switch S2 closes, and R_1, R_2, C_2 are grounded. This makes sure that the T/H output voltage during the hold phase is no longer a function of the input voltage. Through the two phases of the circuit, both S1 and S2 are either at the virtual ground node of the OTA, or at ground, and do not experience any signal swing.

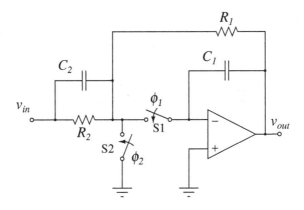

Fig. 5.3: Single-ended T/H circuit, as described in [77].

5.3.1 Charge injection and sampling times

When ϕ_1 goes from high to low and switch S_1 turns OFF, a portion of the charge from the channel of the switch S_1 is dumped onto the capacitor C_1. This creates an error in the value of the charge held in the hold phase during ϕ_2. However, the amount of charged dumped onto C_1 is independent of the input voltage v_{in}, since the total charge in the channel is the same irrespective of v_{in}. Therefore, this charge injection causes a DC offset in the held value. If the circuit is configured to be fully differential, this DC error will be minimized. A random, nominal charge-split component will also appear as a DC offset.

When ϕ_2 goes from high to low and switch S_2 turns OFF, if the impedance of S_2 is small, most of the charge in the channel of the switch S_2 is discharged to ground, and minimal charge is injected into the circuit.

In a practical circuit implementation the clock rise and fall times will be finite and non-zero. This will not make the sampling instants signal-dependent, since the switch S_1 is connected to virtual ground and experiences very small signal swing on either side. The switch S_1 will open and close depending on the voltage between the clock and the virtual ground node and, therefore the switching will have the exact same period as the clock.

5.3.2 Fully-differential implementation

Figure 5.4(a) shows a fully-differential form of the single-ended T/H circuit in Fig. 5.3. At a supply voltage of 0.5 V, to allow for maximum signal swing, the output and input common-mode voltages of the T/H circuit are 0.25 V. Let us assume that we can design an OTA that operates at an input common-mode voltage, v_{cm_i}, of 0.25 V. In the track phase, both sides of the switches S_{1A} and S_{1B} will be at 0.25 V. A clock that swings between 0 and 0.5 V will be insufficient to turn S_{1A} and S_{1B} ON.

Figure 5.4(b) shows a modified fully-differential implementation of the T/H circuit. The fully-differential OTA used is a gate-input OTA [79], which operates at an input common-mode voltage of 0.4 V, and sets an output common-mode voltage of 0.25 V. pMOS switches are used for S_{1A}, S_{1B}, and S_{2A}, S_{2B}. Non-overlapping clocks ϕ_1 and ϕ_2 switch between 0 and 0.5 V, and their complements, $\overline{\phi_1}$ and $\overline{\phi_2}$, are applied to the switches S_{1A}, S_{1B} and S_{2A}, S_{2B}.

In the track phase (when $\overline{\phi_1}$ is low), the inputs of the OTA are a virtual short and are raised to 0.4 V through resistors to V_{DD}, labeled R_{3A} and R_{3B}. This enables operation towards moderate inversion for the OTA input devices. The pMOS switches (S_{1A}, S_{1B}) clocked by ϕ_1 experience a source-gate voltage of 0.4 V, which turns them ON. Their bodies are connected to their gates to further invert them strongly when they are ON; the forward body bias reduces the V_T of the pMOS devices from 0.6 V to 0.4 V. Forward body biasing of the body-source junction has been applied in low-voltage digital circuits [63] with no risk of latch-up for a 0.5 V supply. For S_{1A}, S_{1B} sized as 24 μm wide and 0.25 μm long, their ON resistance is 1.5 kΩ.

In the hold phase (when $\overline{\phi_2}$ is low), a source-gate voltage of 0.5 V is available for the pMOS switches (S_{2A} and S_{2B}). For S_{2A}, S_{2B} sized as 24 μm wide and 0.25 μm long, their ON resistance is 7.2 kΩ. At the same time, the capacitors C_{1A} and C_{1B} hold charge such that the OTA output is held at the final value at the end of the track cycle. This maintains the input common-mode voltage of the OTA at 0.4 V during the hold phase.

5.3.3 Common-mode rejection

The maximum common-mode gain, i.e. the gain from the input common-mode voltage to the output common-mode voltage, of the OTA (to be discussed in Sec. 5.4) in the used 0.25 μm CMOS technology was about 2 dB. During the track phase of the

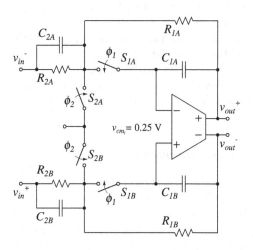

(a)

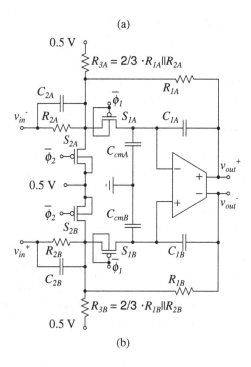

(b)

Fig. 5.4: (a) Fully-differential implementation of Fig. 5.3, assuming the OTA operates at an input common-mode voltage of 0.25 V. It will be a challenge to switch on S_{1A} and S_{1B}. (b) Fully-differential T/H circuit modified for 0.5 V operation, with an OTA that operates at an input common-mode voltage of 0.4 V.

T/H circuit, the common-mode loop-gain is therefore -12 dB (feedback factor of 1/5 because of the resistors – see Fig. 5.4(b)). However, during the hold phase of the T/H circuit, the resistive feedback around the OTA is opened, and the feedback factor, β_H, in the absence of the capacitors C_{cmA}, C_{cmB}, is given by:

$$\beta_H = \frac{C_{1A}}{C_{1A} + C_{in,cm}}$$

where, $C_{in,cm}$ is the effective input capacitance of the gate-input OTA for a common-mode signal. If we assume a common-mode gain for the input stage of 0 dB, the $C_{in,cm}$ is effectively $2C_{gd} + C_{gs}$ of the input nMOS device. Given the device sizes, β_H evaluates to a factor of 2/3. The resulting common-mode loop-gain of -1 dB allows the input common-mode voltage of the OTA to drift. As a result, the high-gain of the OTA reduces, common-mode gain further increases, and the input common-mode voltage further drifts. This leads to potential instability during the hold phase of the T/H, which is unacceptable.

One way to solve this problem is to reduce the OTA common-mode gain. An alternate approach is to reduce β_H, the common-mode feedback factor. Capacitors, denoted as C_{cmA}, C_{cmB} in Fig. 5.4(b), are inserted at the inputs of the OTA; they increase the common-mode input capacitance of the gate-input OTA. For C_{cmA} and C_{cmB} of 1 pF, β_H reduces to 2/5, and the common-mode loop-gain is reduced to -6 dB, and the T/H circuit is now stable under all conditions. In the differential mode, the input capacitance of the OTA is about 1 pF. With the addition of the 1 pF at the input, the phase margin of the system in the track phase decreases. Simulations predicted a phase margin of 80° in the track phase.

5.3.4 Integrated noise

For this analysis we assume the following:

(a) The gain of the OTA is sufficiently large, so that the inputs to the OTA are effectively at the same voltage.
(b) During the track phase, the resistance of the switches S_{1A}, S_{1B} is very small, so that both the drain and the source of the switches are at the same voltage.
(c) The OTA can be represented by a block having a gain of ω_P/s. ω_P is the unity-gain bandwidth of the OTA (in rad/sec).
(d) $R_{1A} = R_{1B} = R_{2A} = R_{2B} = R$, and $C_{1A} = C_{1B} = C_{2A} = C_{2B} = C$. This makes the biasing resistors, $R_{3A} = R_{3B} = R/3$.

The noise in a T/H circuit is the noise in the hold phase of the circuit. During the hold phase, noise observed at the output of the T/H in Fig. 5.4(b) is the sum of the noise contribution of the OTA and the noise sampled onto the hold capacitors, C_{1A} and C_{1B}, at the end of the track phase. The noise contribution of the OTA is small compared to the noise sampled onto C_{1A} and C_{1B}, when the OTA input-referred noise power-spectral density is much lower than $4kTR$. The noise sampled onto the capacitors at the end of the track phase, and a sample of the noise in the track phase, have the same RMS value [80, 81]. To compute the RMS value of a sample of the noise in the track phase, we first compute the noise power-spectral density at the T/H output during the track phase, and then integrate over all frequencies [82, 83].

The noise power-spectral density at the output of the T/H during the track phase is the sum of the individual noise power-spectral density contributions from all of the noise sources. There are six resistors, each contributing noise to the output of the circuit. The other noise sources are the switches, S_{1A}, S_{1B}, and the OTA. If the switch ON resistance is much lower than R, then the dominant noise contribution is from the resistors. If the OTA input-referred noise power-spectral density is much lower than $4kTR$, then the dominant noise contribution is again from the resistors [2]. Fig. 5.5 shows the AC small-signal circuit for the T/H circuit in the track phase.

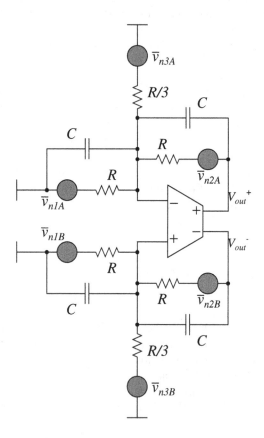

Fig. 5.5: AC small-signal circuit for noise analysis of the T/H circuit in the track phase.

---

[2] If R is 100 kΩ, and the g_m of the input devices in the OTA is of the order of mS, this is indeed true.

The gain, $G_{n1}(\omega)$ from each of the noise sources marked $\overline{v_{n1A}}$, $\overline{v_{n1B}}$ to the respective outputs, and $G_{n2}(\omega)$ from $\overline{v_{n2A}}$, $\overline{v_{2B}}$ to the respective outputs, at any frequency ω, is given by:

$$G_{n1}(\omega) = G_{n2}(\omega) = \frac{1}{1 + j\omega RC + \left(\frac{-2\omega^2 RC + 5j\omega}{\omega_P}\right)}$$

The gain, $G_{n3}(\omega)$ from the noise sources marked $\overline{v_{n3A}}$, $\overline{v_{n3B}}$ to the differential output, is given by:

$$G_{n3}(\omega) = 3 \cdot G_{n1}(\omega)$$

The total power-spectral density at the differential output of the T/H circuit is the sum of the mean-square noise density contributions of each of the individual resistors at the output nodes. Assuming the OTA unity-gain bandwidth, ω_P, is large compared to $1/(RC)$ and neglecting higher order terms, the noise power-spectral density at the output is given by:

$$\overline{S_n(f)} = 2|G_{n1}(f)|^2 \cdot 4kTR + 2|G_{n2}(f)|^2 \cdot 4kTR$$
$$+ 2|G_{n3}(f)|^2 \cdot 4/3 \cdot kTR$$

$$\approx \frac{2 \cdot 4kTR + 2 \cdot 4kTR + 2 \cdot 3 \cdot 4kTR}{1 + 4\pi^2 f^2 (R^2 C^2 + \frac{6RC}{\omega_P})}$$

The power-spectral density at any frequency, f, at the differential output of the T/H circuit, is now given by:

$$\overline{v_n^2(f)} = \frac{40kTR}{1 + 4\pi^2 f^2 (R^2 C^2 + \frac{6RC}{\omega_P})} \tag{5.2}$$

Integrating (5.2) over all frequencies gives us the total integrated mean-square differential output noise voltage during the track phase:

$$\int_0^\infty \overline{v_n^2(f)} \, df = 10 \, \frac{kT}{C} \, \frac{1}{\sqrt{1 + \frac{6}{\omega_P RC}}} \tag{5.3}$$

For $\omega_P \gg 1/RC$, the mean-square noise voltage approximates to $10kT/C$. Of this total integrated mean-square noise voltage of $10kT/C$, a contribution of kT/C can be traced back to each of the four resistors of value R, and a contribution of $3kT/C$ can be traced back to each of the two biasing resistors, R_{3A}, R_{3B}, of value $R/3$. Choosing C_{1A}, C_{1B} and C_{2A}, C_{2B} equal to 1 pF, R_{1A}, R_{1B}, and R_{2A}, R_{2B} equal to 100 kΩ, and an OTA unity-gain bandwidth of 20 MHz results in a total integrated equivalent input noise of 200 μV_{RMS}. A pseudo-differential arrangement of the basic T/H circuit in Fig. 5.1 has an RMS noise voltage of $\sqrt{2kT/C}$, which is $\sqrt{5}$ times smaller than the noise of the T/H circuit in Fig. 5.4(b).

5.3.5 Track-and-hold test strategy

Two identical T/H circuits can be cascaded, to realize a sample-and-hold circuit. The first T/H is referred to as the master T/H, and the second one as the slave. At the end of the hold phase of the master T/H, the slave T/H is switched from track to hold. The overall system is transparent only during the short period of time when both the master and the slave T/H are in track mode. Both the T/H circuits are sampled at the same clock frequency. The required clock waveforms for the master and slave T/H circuits are shown in Fig. 5.6.

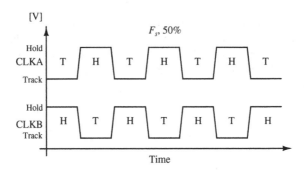

Fig. 5.6: Required clock waveforms for $\overline{\phi_1}$ of the master (CLKA) and slave (CLKB) T/H circuits for sample-and-hold operation.

Alternatively, two T/H circuits can be cascaded and configured as in Fig. 5.7 [84], to test the high-frequency distortion of a single T/H circuit. In this re-sampler setup, the output of the master T/H is re-sampled by a second, identical, slave T/H. The slave T/H is clocked at half the frequency of the master T/H. The clock waveforms are shown in Fig. 5.7(b). If the input signal is at a frequency close to the Nyquist rate, say $F_s/2 - \Delta f$, where F_s is the master T/H clock frequency, then the final output of the slave T/H is at a frequency Δf. This simplifies the distortion measurement.

5.4 Design details and measurement results

5.4.1 Gate-input OTA

The gate-input OTA was designed [79] for a gain-bandwidth of 20 MHz, a DC gain of 55 dB and a slew rate of 6.6 V/μsec (Fig. 5.8). This allows an SNDR of 60 dB up to the Nyquist frequency for 1 Msps sampling. To improve the OTA gain-bandwidth product, forward body biasing was used to reduce the transistor V_Ts [50]. The OTA schematic is shown in Fig. 5.8, and the sizes of the different devices and passives in the OTA are tabulated in Table 5.1. The gain and phase response of the OTA for a load capacitance of 20 pF to ground on each output (10 pF differential), are shown

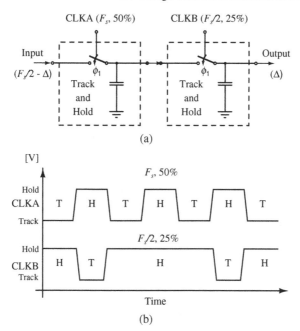

Fig. 5.7: (a) Two T/H circuits cascaded, mixing down the output to a low frequency. (b) Required clock waveforms for the two T/H circuits.

in Fig. 5.9. Biasing techniques as in [79] were implemented for automatic on-chip biasing and gain enhancement.

5.4.2 Switches

The pMOS switches S1 and S2 are sized as 24 μm wide and 0.25 μm long. When ON, a V_{SG} of 0.4 V is available to S1, while a V_{SG} of 0.5 V is available to S2. To reduce the ON resistance of S1, the body of S1 is switched along with the gate, as indicated in Fig. 5.4(b). The resulting ON resistance of S1 is 1.5 kΩ, and that of S2 is about 7.2 kΩ. As discussed earlier in Section 5.3.1, charge injected from the switches onto the hold capacitor, C_1, is independent of the input signal, and thus creates a DC offset. Differentially, in the absence of mismatches, this DC offset is canceled.

5.4.3 Clock generation

For operation of the T/H circuit, non-overlapping clock waveforms, $\overline{\phi_1}$ and $\overline{\phi_2}$ are required. The circuit in Fig. 5.10 generates two non-overlapping clock waveforms from an input clock ([47], pp. 397-398). The two back-to-back NOR gates form an S-R latch. The S-R latch does not allow both outputs to be high at the same time. If

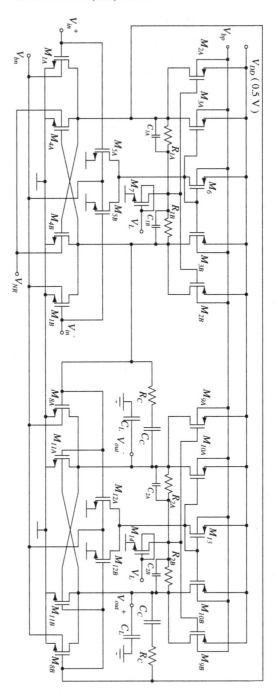

Fig. 5.8: Gate-input OTA [79] designed for a 0.25 μm CMOS process, and a unity-gain bandwidth of 20 MHz. Component values are given in Table 5.1.

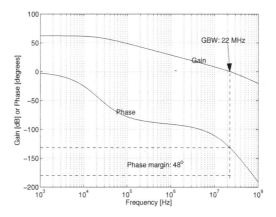

Fig. 5.9: Gain and phase response of the gate-input OTA designed for the T/H circuit.

Table 5.1: Component sizes and values for the gate-input OTA designed in a 0.25 μm CMOS technology (Fig. 5.8).

First stage			Second stage		
Transistors	W [μm]	L [μm]	Transistors	W [μm]	L [μm]
M_{1A}, M_{1B}	2000	0.25	M_{8A}, M_{8B}	2000	0.25
M_{2A}, M_{2B}	200	0.25	M_{9A}, M_{9B}	100	0.25
M_{3A}, M_{3B}	200	0.25	M_{10A}, M_{10B}	200	0.25
M_{4A}, M_{4B}	200	0.25	M_{11A}, M_{11B}	200	0.25
M_{5A}, M_{5B}	200	0.25	M_{12A}, M_{12A}	200	0.25
M_6	40	0.25	M_{13}	40	0.25
M_7	160	0.25	M_{14}	80	0.25
Resistors and Capacitors					
R_{1A}, R_{1B}	50 kΩ		R_{2A}, R_{2B}	50 kΩ	
C_{1A}, C_{1B}	1 pF		C_{2A}, C_{2B}	1 pF	
			R_C	600 Ω	
			C_C	15 pF	

T is the period of the input clock of duty cycle 0.5, Δ_I is the delay in an inverter, and Δ_N is the delay in a NOR gate, then the duty cycles of ϕ_1 and ϕ_2, η_1 and η_2, are given by:

$$\eta_1 = 0.5 - \frac{\Delta_N + \Delta_I}{T}$$

$$\eta_2 = 0.5 - \frac{\Delta_N - \Delta_I}{T}$$

The delay between the falling edge of ϕ_1 and the rising edge of ϕ_2, and vice-versa, is Δ_N.

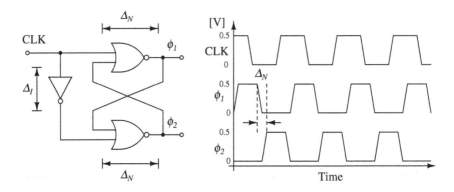

Fig. 5.10: Non-overlapping clock generating circuit. The bodies of the pMOS devices in the inverter and NOR gate are connected to the ground voltage, while the bodies of the nMOS devices are connected to V_{DD}.

The clock generating circuit is not perfectly symmetric and the duty-cycles in ϕ_1 and ϕ_2 are not equal. The inverter, required to generate the complement of the input clock, is the most critical from a symmetry point of view. The inverter is therefore sized to have a fanout of 16, and the delay Δ_I is minimized.

For the clock generating circuit to be functional at a given frequency, both η_1 and η_2 have to be greater than zero. At the low power supply voltage of 0.5 V, this is of much concern, since the devices are not in strong inversion and are inherently slow. The bodies of nMOS and pMOS devices are swapped [63] to improve the inversion of the devices and reduce the delays in the inverter and NOR gate. In the worst case, (at the slowest simulation corner and at 0°C), for a 1 MHz clock, η_1 and η_2 are simulated to be about 0.48 and 0.50 respectively, and the minimum delay between the falling edge of ϕ_1 and rising edge of ϕ_2, or vice-versa, is about 20 nsec.

5.4.4 Prototype chip

The circuit was fabricated in the CMOS part of a 0.25-μm BiCMOS process, using deep-nwell devices, high-resistivity poly resistors and MIM capacitors. The prototype (Fig. 5.11) contained two cascaded T/H circuits, two non-overlapping clock generators, and OTA biasing circuits. Each core T/H circuit had an active area of 0.4 mm×0.4 mm. Inputs and outputs of both the T/H circuits were accessible through bonding pads.

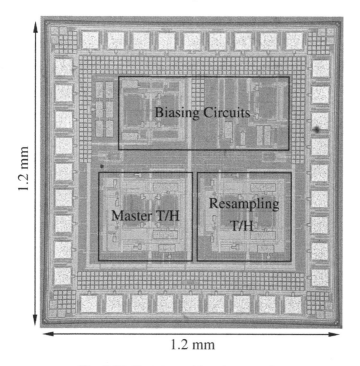

Fig. 5.11: Prototype chip micro-graph.

5.4.5 Simulated performance

The prototype chip is simulated over all process corners and temperatures. Fig. 5.12 shows the time-domain re-sampled output voltage waveform. The input voltage is at a frequency of 496.09375 kHz (127/256 × 1 MHz), and the clock frequency is at 1 MHz. The re-sampled signal has, therefore, a fundamental frequency of 3.90625 kHz. Fig. 5.13 shows a simulation of the signal to noise and distortion ratio (SNDR) for different input amplitudes, on the extracted layout. The frequency of the input signal, for this simulation, was also chosen as 496.09375 kHz – very close to the Nyquist frequency, so that the T/H circuit undergoes maximum slewing. The maximum simulated SNDR is 62 dB, which is better than the design target of 60 dB. Nominally, at a 0.5 V power supply voltage, each of the T/H circuits has an expected current consumption of 400 μA.

5.4.6 Measured performance

One T/H nominally consumes 0.6 mA of current from the 0.5 V power supply. The measured integrated input noise for each T/H circuit is 188 μV_{RMS}. The integrated input noise was obtained by keeping the first T/H in track mode and integrating the noise spectrum measured at the output of the first T/H. Fig. 5.15 shows the measured

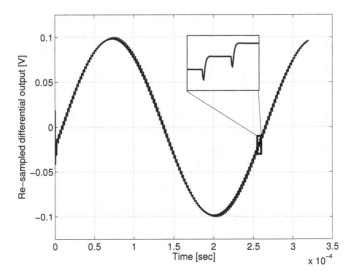

Fig. 5.12: Simulated re-sampled output voltage waveform for an input frequency of 496.09375 kHz.

pedestal and droop rate of the T/H when clocked at 1 kHz. In track mode, the circuit has a slew-rate of 5 V/μsec, and a bandwidth of 3.9 MHz, which corresponds to an estimated OTA bandwidth of 15.6 MHz.

The two T/H circuits were cascaded in the configuration shown in Fig. 5.7. The first T/H circuit is clocked at 1 Msps, while the second re-sampling T/H is clocked at 0.5 Msps with 25% duty cycle. Fig. 5.13 shows the SNDR of the T/H circuit for varying input amplitudes and different input frequencies. For this chapter, SNDR is defined as the ratio of the signal power at the output, to the total of the output power at the harmonics of the input frequency, and output noise. Fig. 5.14 shows the re-sampled differential output at 25 kHz for a 200 mVpp 475 kHz differential input.

Performance metrics of the T/H are summarized, and compared with simulations, in Table 5.2. Compared to simulations, the track-mode bandwidth and the slew-rate are lower. This explains the slight improvement in circuit noise and the slight degradation in SNDR at higher input frequencies compared to simulations.

5.5 Conclusion

In this chapter, a fully-integrated ultra-low voltage track-and-hold circuit has been designed and fabricated in the CMOS part of a 0.25-μm BiCMOS technology, using standard 0.6-V V_T devices. The design is a true low-voltage design - all nodes in the circuit are within the power rails at all times. The T/H circuit has a dynamic range of 60 dB at a sampling clock frequency of 1 MHz, and a power consumption from a 0.5-V power supply of 300 μW. The only other sampled circuit at 0.5 V published in

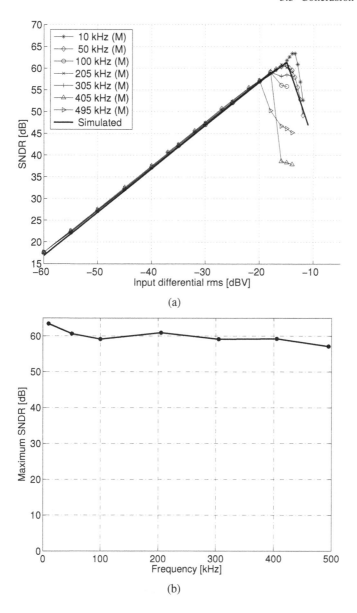

Fig. 5.13: (a) Measured signal-to-noise-and-distortion ratio at different input signal frequencies. (b) Measured maximum signal-to-noise-and-distortion ratio as a function of input signal frequency.

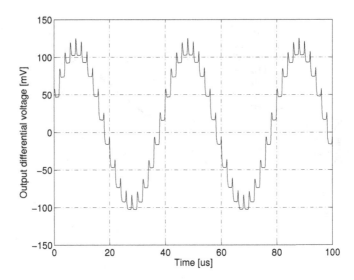

Fig. 5.14: Re-sampled 25 kHz (differential) output waveform for a 200 mVpp (differential) input at 475 kHz.

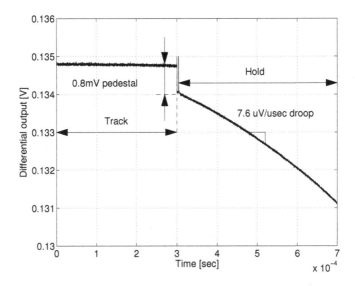

Fig. 5.15: Measured pedestal voltage and droop rate of the T/H circuit.

Table 5.2: Performance of the master T/H circuit

	Measured	Simulated
Power supply	0.5 V	0.5 V
Current consumption	600 μA	600 μA
Chip area	0.4 mm$\times$0.4 mm	–
Sampling rate	1 Msps	1 Msps
Differential input referred noise	188 μV$_{RMS}$	200 μV$_{RMS}$
Peak SNDR @ f_{in} = 50 kHz	60 dB	62 dB
Peak SNDR @ f_{in} = 495 kHz	57 dB	60 dB
Hold mode droop rate on diff. output	7.6 μV/μsec	–
Pedestal on differential output	0.8 mV	–
Track mode bandwidth	3.9 MHz	5 MHz

open literature [40] to our knowledge, reports 43 dB of SNDR for a 4.1 kHz sampling clock frequency. The topology presented here also holds the promise of much faster operation using scaled devices in nano-scale CMOS technologies, while being fully compatible with the required ultra-low supply voltages.

6

A 0.5 V Continuous-Time $\Sigma\Delta$ Modulator[1]

6.1 Introduction

In this chapter an audio-band continuous-time (CT) $\Sigma\Delta$ modulator is presented as another example of true low voltage design without using low threshold devices or internal voltage boosting. A $\Sigma\Delta$ modulator has a higher level of design complexity than the circuit examples presented in the previous chapters. It requires a clocked comparator and feedback digital-to-analog converters (DAC) in addition to a loop filter of either continuous-time or discrete-time type.

In recent years, a number of discrete-time low-voltage $\Sigma\Delta$ modulators have been reported. Most of them are implemented using a switched-opamp or related topology [37, 40, 42, 85, 86]. Low-voltage continuous-time modulators with a supply voltage down to 0.9 V have been reported in [87, 88].

We have chosen to realize a continuous-time modulator for three reasons. First, it requires less switches than its discrete-time counter part. Implementing switches at a low supply voltage without using low threshold devices or clock boosting is well-known to be hard. Second, a continuous-time modulator has more relaxed requirement on the amplifier's bandwidth than its discrete-time counterpart. Finally, a continuous-time modulator carries the advantage of having an inherent anti-aliasing function. The ultra-low voltage continuous-time modulator reported in this chapter uses a newly proposed return-to-open (RTO) modulator architecture in combination with body-input gate-clocked circuits to operate with a power supply voltage down to 0.5 V [58].

The rest of this chapter is organized as follows. Section 6.2 discusses the challenges of implementing, at low supply voltages, the return-to-zero signaling scheme, which is essential in the feedback digital-to-analog converter in a low-distortion CT modulator. The RTO technique is presented as a solution to this challenge. The im-

---

[1] ©2007 IEEE. Portions reprinted, with permission, from K. Pun, S. Chatterjee, P. Kinget, "A 0.5-V 74-dB SNDR 25-kHz Continuous-Time Delta-Sigma Modulator with a Return-to-Open DAC", *IEEE Journal of Solid State Circuits*, vol. 42, no. 3, pp.496-507, March 2007.

provement on the modulator thermal noise performance by using the RTO technique is also analyzed. Section 6.3 shows the details of the modulator design. Section 6.4 presents the 0.5 V building block circuits. Section 6.5 has the experimental results and is followed by conclusions in section 6.6.

6.2 Return-to-Open DAC

It is well known that the return-to-zero (RZ) signaling scheme is required in the feedback DAC of a CT $\Sigma\Delta$ modulator for low distortion [89]. Fig. 6.1 shows a typical active-RC integrator stage in a 1-bit CT modulator. If non-return-to-zero (NRZ) signaling is used, data dependent transients exist at the DAC's output node due to its parasitic node capacitance. This data-dependent transient causes inter-symbol-interference (ISI), which degrades the linearity of the modulator . Even a differential circuit implementation, which has symmetric DAC rising and falling edges, cannot remove the ISI. In the illustrative DAC waveform in Fig. 6.1, the rising and falling edges are symmetrical. Despite this symmetry, the area for the first symbol "1" and that of the third symbol "1" are different and dependent on their previous symbol value. In contrast, by using RZ signaling, the DAC's output resets to a constant DC level before the next input comes in, removes the data dependent transient and thus improves the linearity of the modulator.

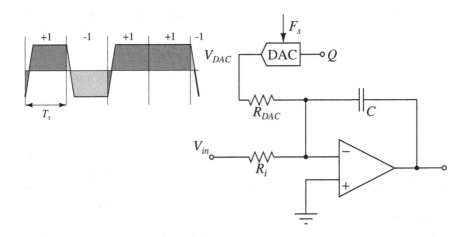

Fig. 6.1: A typical integrator stage of an active RC $\Sigma\Delta$ modulator with a NRZ DAC that has symmetrical rising and fall edges.

A big challenge brought by the RZ scheme for a low supply voltage implementation is the need of switches operating at $V_{DD}/2$. In low voltage designs, a differential implementation of the integrator (see Fig. 6.2) is almost mandatory for its larger signal swing. During the active phase, the positive DAC connects to V_{DD} and the negative DAC connects to 0 V when the DAC input is "1", or the other way round when the DAC input is "-1". At this time, the amplifier's input common-mode (CM) level is $V_{DD}/2$ in the typical case that the CM level of V_{in} is at $V_{DD}/2$ for maximal signal swing. During the inactive phase[2], the positive and negative DACs must connect to the same voltage level, V_{CM}, in order to produce a differential signal of "0". Though in theory this V_{CM} can be of any value if the differential amplifier has perfect CM rejection ability, it is typically set at $V_{DD}/2$ to avoid over exercising the amplifier's input CM level. But with a supply voltage of 0.5 V and a CM signal level of 0.25 V there is only 0.25 V available to turn on the switches, which is often not sufficient.

We propose to remove the problem switches connected to VDD/2 as shown in Fig. 6.3. The remaining circuit still performs the RZ function and the circuit is referred to as a return-to-open (RTO) DAC . Consider the case when the DAC input Q is "1". During the active phase, the positive DAC connects its output, V_2, to V_{DD} and the negative DAC connects its output, V_1, to 0 V. So far, the operation is the same as that of a conventional RZ DAC. But when the RTO DACs enter the inactive phase, their outputs are not connected to any reference voltage, but are open-circuited and left floating. Now the resistors R_{DAC} are open circuited at one end and the currents flowing from them to the integrating capacitor are zero. This means that the DACs are essentially returned to "zero" during the inactive phase.

In the inactive phase, the outputs of the DACs are connected to the amplifier's input terminals through the resistors R_{DAC}, and therefore their potential will be pulled to the amplifier's input CM level as illustrated in Fig. 6.3(b). The amplifier's input CM level at this time is determined by the CM level of V_{in}, or the output of the previous integrator stage. Both the CM levels are typically set at $V_{DD}/2$ for maximal signal swing. Thus the DAC's outputs are reset to $V_{DD}/2$ in every RZ interval. Any signal-dependent charges that may be stored in the parasitic capacitors at the DAC's output nodes will be cleared in the inactive phase. ISI is therefore eliminated without using any switch operating at $V_{DD}/2$.

6.2.1 Similar DAC concepts

DAC circuits similar to the RTO concept for CT $\Sigma\Delta$ modulators have been reported in [90–92]. The major difference is that the RTO DAC intentionally removes switches operating at any level between V_{DD} and ground, while in all other DAC circuits [90–92] a switch operating at a CM level between V_{DD} and ground is inevitable, making them less suitable for ultra-low voltage operation.

In [90], a 1.5-bit DAC with RZ output is used for a CT $\Sigma\Delta$ modulator. The DAC has three output states: +1, 0 and -1. In the 0-state, the DAC injects zero differential current to the integrator, similar to the RTO DAC concept. However, the 0-state is not implemented by leaving DAC resistors open at one end as done here. Instead,

---

[2] In this chapter, we define "inactive phase" as the time during which the DAC output is differentially zero, and "active phase" as the time during which the DAC output is differentially ± 1.

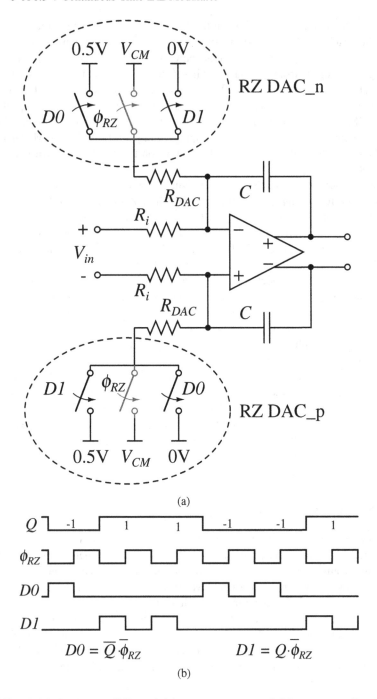

(a)

(b)

Fig. 6.2: (a) RZ DAC in a differential integrator stage and (b) corresponding clock signals and data. Q is the digital output of the 1-bit comparator of the modulator.

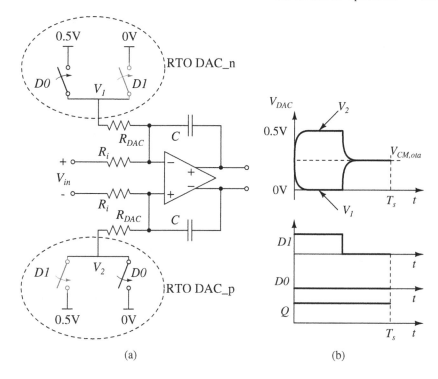

Fig. 6.3: (a) Return-to-open DAC concept, and (b) corresponding DAC output waveforms and data and clock signals.

the positive and negative DAC resistors are short-circuited by a switch. This switch is necessary for removing the effect of amplifier offset on the linearity of the feedback DAC, which is crucial in a multi-bit modulator (including the 1.5-bit case). The switch then must be able to operate at the amplifier's input CM level, making the DAC not suitable for ultra-low voltage operation.

In [91], the switched-resistor-capacitor (SRC) feedback DAC is used for a CT $\Sigma\Delta$ modulator. The purpose of using SRC DAC is to reduce the modulator's sensitivity to clock jitter, which is achieved by utilizing the exponential decay property of the RC discharging circuit. The behavior of the SRC DAC during the DAC capacitors' discharging phase (its active phase) is the same as that of the RTO DAC in the inactive phase. However, unlike the RTO DAC, the SCR DAC must disconnect its output from the integrator during the capacitor charging phase for proper operation. Consequently, a switch operating at the amplifier's input CM level is inevitable as shown in the SRC DAC circuit in [91].

In [92], a switched-capacitor (SC) feedback DAC is used for a CT $\Sigma\Delta$ modulator with the same purpose of reducing the sensitivity to clock jitter. Taking the finite switch on resistance into account, the SC DAC is the same as the SRC DAC. The SC DAC, again, requires switches operating at the amplifier's input CM level.

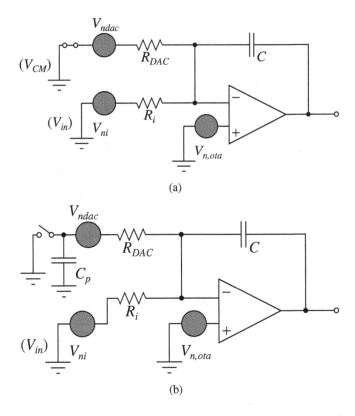

(a)

(b)

Fig. 6.4: Noise models for the integrator employing (a) the conventional RZ DAC, and (b) RTO DAC, both in the inactive phase.

6.2.2 Noise improvements by RTO DAC

The RTO DAC has an additional advantage over the conventional RZ DAC with respect to the noise performance. During the active phase, the RTO DAC and the conventional RZ DAC have identical circuitry and thus identical noise performance. But during the inactive phase, their noise performance is very different. Fig. 6.4(a) and Fig. 6.4(b) depict the noise models for the modulator first stage employing the conventional RZ DAC and the RTO DAC respectively. The squares of their equivalent input noise voltages are given below in (6.1) and (6.2) respectively:

$$\overline{v_{neq,RZ}^2} = \overline{v_{ni}^2} + \overline{v_{ndac}^2}(\frac{R_i}{R_{dac}})^2 + \overline{v_{n,ota}^2}(\frac{R_i}{R_i||R_{dac}})^2 \tag{6.1}$$

$$\overline{v_{neq,RTO}^2} = \overline{v_{ni}^2} + \overline{v_{ndac}^2}\left|\frac{R_i}{R_{dac} + 1/j\omega C_p}\right|^2 + \overline{v_{n,ota}^2}\left|\frac{R_i}{R_i||(R_{dac} + 1/j\omega C_p)}\right|^2 \tag{6.2}$$

where $\overline{v_{ni}^2} = 4kTR_i \cdot BW$, $\overline{v_{ndac}^2} = 4kTR_{dac} \cdot BW$, with BW the signal bandwidth, $\overline{v_{n,ota}^2}$ is the square of the amplifier's total input referred noise voltage within BW, and C_p is the parasitic capacitor at DAC's output node. At low frequencies ($\omega \ll 1/R_{dac}C_p$), (6.2) reduces to:

$$\overline{v_{neq,RTO}^2} = \overline{v_{ni}^2} + \overline{v_{ndac}^2} \times 0 + \overline{v_{n,ota}^2} \times 1 \qquad (6.3)$$

As can be seen from (6.3), R_{dac} does not contribute noise to the modulator at low frequencies, because R_{dac} is open-circuited at one end, preventing low frequency noise current from flowing into the integrating capacitor. Moreover, the equivalent input noise voltage of the op-amp is not amplified when referred to the input for the same reason that R_{dac} is floating at one end. The assumption for (6.3) is valid for this design since the highest signal frequency is 25kHz, and C_p is mainly contributed by the gate-drain capacitances of the output stage transistors of the RTO DAC (see Section 6.4.1), which are very small.

The noise expressions of equation (6.1)-(6.3) are for the inactive phase. For the active phase, equation (6.1) applies to both the conventional RZ DAC and the RTO DAC. With a duty cycle of δ ($0 < \delta < 1$), the RTO DAC has an average input noise power of:

$$\overline{v_{neq,RTO}^2} = \overline{v_{ni}^2} + \overline{v_{ndac}^2}\delta\left(\frac{R_i}{R_{dac}}\right)^2 + \overline{v_{n,ota}^2}\left[\delta\left(\frac{R_i}{R_i||R_{dac}}\right)^2 + (1-\delta)\right] \qquad (6.4)$$

6.2.3 Return to Zero timing

A minor drawback of the RTO DAC is a slightly longer return-to-zero transition time. Referring to Fig. 6.3(b), in the inactive phase, the output of the RTO DAC is exponentially discharged to the "0" level with a time constant of $R_{dac}C_p$, where C_p is the parasitic capacitor at DAC's output node. The conventional RZ DAC has a smaller discharging time constant $R_{ON}C_p$, where R_{ON} is the on-resistance of the switch which is usually much smaller than R_{dac}. Attention must be paid to this prolonged return-to-zero falling time when choosing the value of R_{dac} for the modulator. Finally, the non-ideality of the exponential falling edge could be taken into account in the synthesis of the CT modulator [93], if desired.

6.3 Split RTO DAC Modulator Architecture

A straightforward way of applying the RTO DAC in a CT modulator is to use one RTO DAC to drive all the feedback resistors. However such an implementation suffers from unwanted inter-stage signal coupling unless the modulator has only one feedback path. When the DAC output floats, signals from different stages can couple to each other through the floating node. These unwanted signal paths alter the loop transfer function and deteriorate the modulator's performance. This inter-stage signal coupling is removed by splitting the DAC into several DACs, one for each feedback path, as shown in Fig. 6.5. Now, the feedback resistors are isolated from each other, the parasitic signal paths are cut and the loop transfer function of the modulator remains intact. Although the number of DACs is increased, the total resistive load of the DACs remains the same as in the single DAC case. Each of the DAC sizes can be appropriately sized so that splitting the DAC does not have a penalty on the size or power.

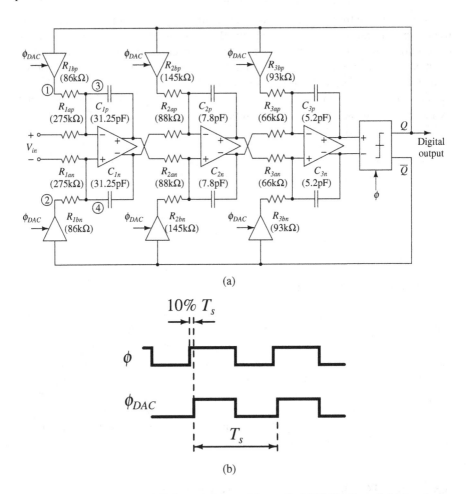

(a)

(b)

Fig. 6.5: (a) The proposed $\Sigma\Delta$ modulator with a split RTO DAC, and (b) its clock signals.

6.3.1 Modulator clocking

At a supply of 0.5 V, it takes a longer time for the comparator's output to settle. This settling transient is signal dependent, which could deteriorate the modulator's linearity

To suppress this potential source of distortion, the clock for the DACs (ϕ_{DAC}) is delayed from the clock for the comparator (ϕ) by 10% of a clock cycle to allow the comparator outputs to fully settle before the DACs become active. The duty cycle of ϕ_{DAC} was chosen as 50%.

6.3.2 Modulator design

To demonstrate this ultra-low voltage modulator architecture, a third-order 1-bit CT modulator was designed. The signal bandwidth is 25 kHz and the sampling frequency is 3.2 MHz, which corresponds to an over-sampling ratio (OSR) of 64. The OSR was chosen as a sufficiently large integer power of two to guarantee that the modulator can still provide a satisfactory SNDR when there is a large RC time-constant variation.

Figure 6.6 shows the modulator coefficients, which were obtained by an impulse-invariant transformation [94] of a discrete-time modulator prototype designed with a Matlab^{TM} toolbox [95]. The DAC output waveform was assumed to be rectangular and the 50% duty cycle and the 10% delay of the DAC have been taken into account in the modulator design. The model showed in Fig. 6.6 has maximum stable input amplitude of approximately 0 dB with respect to the full scale value of the feedback DAC, as revealed by behavioral simulations.

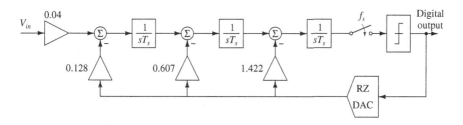

Fig. 6.6: Modulator coefficients (f_s = 3.2 MHz, T_s=312.5 ns).

Figure 6.7 shows the behavioral simulation results of the modulator. The SNDR is plotted for RC product variation (i.e., the variation of the value of T_s in the $1/sT_s$ blocks in Fig. 6.6) from -30% to +50% under different ϕ_{DAC} delay values. The results are obtained with V_{in} fixed at -2.5 dB (slightly lower than maximum stable point of 0 dB) with respect to the reference voltage of the feedback DAC. Under the nominal 10% T_s ϕ_{DAC} delay, the SNDR is better than 77 dB for the whole RC product variation range. Excluding the unstable case of 15%T_s ϕ_{DAC} delay at -30% RC variation, the SNDR is over 76 dB for all cases. As a result, frequency tuning for the loop filter was not implemented. In the actual design, the RC values have been intentionally increased by 10% from their nominal values so that the design falls at the center of the permissible RC variation range. With this change, the modulator can achieve better than 77 dB SNDR for a ±40% RC variation under the nominal 10%T_s ϕ_{DAC} delay.

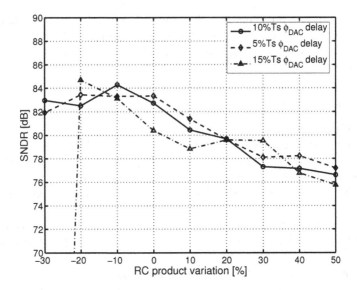

Fig. 6.7: Behavioral simulation results: SNDR versus RC product variation at different ϕ_{DAC} delay, with V_{in} at -2.5 dB with respect to the reference voltage of the feedback DAC.

6.3.3 Values of R and C

The values of the resistors and capacitors of the modulator are chosen based on constraints of noise, chip size and the amplifier's driving capability. For the noise, only those from the first integrator is important as noises from the subsequent stages are suppressed by the high gain of the integrator in the signal bandwidth. Fig.6.8(a) shows the modulator's signal-to-thermal-noise ratio (STNR) for input resistance varying from $50k\Omega$ to $500k\Omega$. The STNR here is defined as the ratio of the maximum input signal power (with input having the maximum amplitude of $0.5V_{diff}$) to the power of the input referred noise of the first stage of the modulator, excluding the noise contributed by the amplifier. The STNR for the modulator with a RTO DAC (based on (6.4)) and with a conventional RZ DAC are plotted as solid curve and dotted curve respectively in Fig.6.8(a). One can clearly see that the RTO improves the noise performance by more than 2 dB over the whole range.

Figure 6.8(a) shows that a smaller input resistance leads to a better STNR. However, a smaller resistance results in a larger integrating capacitance. From the modulator architecture shown in Fig.6.6, we have the following relationship between R_i and C_1 for the first active RC integrator:

$$\frac{1}{R_i C_1} = \frac{0.04}{T_s}, \qquad \text{where } T_s = 312.5ns \qquad (6.5)$$

This relationship is plotted in Fig.6.8(b). Considering the STNR and the capacitor size, the input resistance is chosen as 275 kΩ, and the corresponding capacitance for

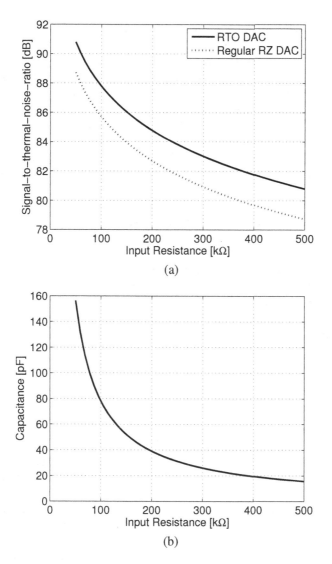

Fig. 6.8: (a) Effect of the input resistance on the modulator's signal-to-thermal-noise ratio, defined at $V_{in} = 0.5V_{diff}$; (b) Capacitance of the first integrator versus input resistance.

the first integrator is 31.25 pF. The resulted STNR is 83 dB, which is not a limiting factor compared to the SNDR shown in Fig.6.7. Moreover, the STNR varies less then 1 dB for the resistance varying by $\pm 20\%$.

The R and C values for the second and third stages are mainly determined by equating the dynamic ranges at the outputs of different stages and by considering the driving capability of the amplifiers. The resulted R and C values are shown in Fig. 6.5. The smallest resistance in the modulator is 66 kΩ, so that the operational transconductance amplifiers (OTAs) can be used without significant DC gain degradation.

In the physical implementation, Metal-Metal (MIM) capacitors and high resistivity poly resistors were used for their low voltage and temperature coefficients.

6.4 Building Block Circuits for 0.5 V Supply

In this section the key building blocks for the modulator implementation are described. A prototype modulator was implemented on a 0.18 μm CMOS technology with threshold voltages of $|V_{TP}| \approx V_{TN} \approx 0.5$ V for zero body-source voltage. Extensive use of the transistor's body terminal has been made for biasing or signal inputs. As long as voltage transients on the power supply are properly controlled, the risk for latch-up in the circuit is non-existent when operating from a 0.5 V power supply [63].

6.4.1 RTO DAC Circuit

Figure 6.9 shows the circuit implementation of the RTO DAC. The DAC consists of two single-phase clocked inverters connected in series and an internal reset switch M_4. All the body terminals are connected to $V_{DD}/2$ (i.e. Vbn=Vbp=$V_{DD}/2$) to lower the threshold voltage of the devices and thus, to ensure proper logic gate operation at the required clock speed (3.2 MHz). The DAC takes the comparator's output Q as its input and uses two reference voltages V_{REFP}=0.5 V and $V_{REFN} = 0$ V.

During the active phase, ϕ_{DAC} is HIGH, and M_3 and M_7 are ON and M_4 is OFF. The two inverters (consisting of M_1-M_2 and M_5-M_6 respectively) are enabled, passing the value of Q to the output at node 2. During the inactive phase, ϕ_{DAC} goes LOW, and M_3 and M_7 are OFF. M_4 turns ON and connects internal node 1 to V_{REFP} so M_5 turns OFF. There is no path available for the output node 2 to rise or fall. So the DAC returns to the "OPEN" state when ϕ_{DAC} is LOW. So far the operation of the RTO DAC is similar to the operation of a tri-state digital buffer [96, 97].

The output will be floating for ϕ_{DAC} LOW (even without transistor M_4). When the clock ϕ of the comparator is LOW, its output Q becomes invalid (as will be explained in the next subsection) and contains signal-dependent transients. Without transistor M_4, this signal dependent transient would be coupled from Q to node 1 then to the output node 2 through C_{gd} of M_1, M_2, and M_5, M_6 and consequently deteriorate the modulator's linearity. Through transistor M_4 the signal-dependent charges injected from Q to node 1 are absorbed to the reference V_{REFP} and do

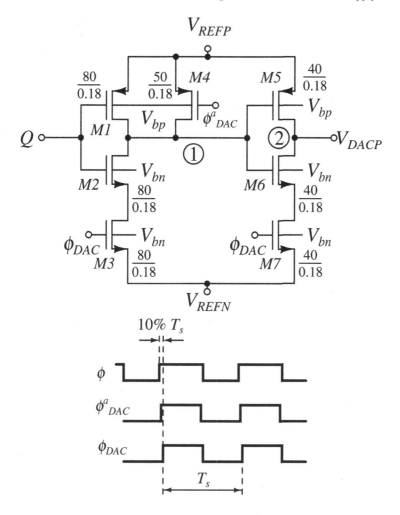

Fig. 6.9: The RTO DAC circuit and its clocks. Vbn=Vbp=$V_{DD}/2$. ϕ^a_{DAC} is a slightly advanced version of ϕ_{DAC}.

not appear at the output. M_4 ensures the charge-injection into node 2 is signal-independent thus eliminating this source of signal dependent errors and hence distortion.

To split the RTO DAC for the modulator, only the second stage consisting of M_5-M_7 is split for different feedback paths. In this design, the second stage transistor sizes for all the feedback paths are the same and are shown in Fig. 6.9.

The simulated voltage waveform at the output (node 1 in Fig. 6.5(a)), of the first positive DAC of the modulator is shown Fig. 6.10(a). The dotted curve is the voltage waveform at the first OTA's input, node 3 in Fig. 6.5(a). As predicted, the DAC output is exponentially discharged to the OTA's input level during the inactive phase. The resulting waveforms for "1" or "-1" are exactly the same, independent of the previous symbol. There is no memory effect and thus no ISI. The waveform for a "1" is not symmetric to that for a "-1" but, this asymmetry does not affect the linearity of a 1-bit modulator since it only affects the OTA's input CM level. As can be seen from Fig. 6.10(a), the OTA's input CM level slightly deviates from 0.25 V to compensate for this asymmetry.

The glitch observed in the waveform for a "1" in Fig. 6.10(a) is caused by the reset transistor M_4. In the actual circuit, the clock ϕ^a_{DAC} for M_4 is a slightly advanced version of the clock ϕ_{DAC} for M_3 and M_7. Switch M_4 turns ON earlier than the moment when switch M_7 turns OFF, so that the signal-dependent charges from input are guaranteed not to appear at the DAC output. A side effect is that the DAC output tends to connect to V_{REFN} shortly even if the input is "1", leading to the glitch observed for "1". This side effect has a benefit of sharpening the return-to-zero transition for the differential DAC. By subtracting the negative DAC's output from the positive DAC's output, the differential DAC output waveform is obtained as shown in Fig. 6.10(b). The rising/falling edge of a symbol at the beginning of each RZ interval is sharpened in the differential output for the positive and negative DACs having similar waveforms in that interval. The differential DAC waveform becomes very close to the ideal case of rectangular waveform, based on which the CT modulator coefficients were designed.

6.4.2 Comparator

At a 0.5 V supply, a regular gate-input comparator does not work when the inputs to the comparator have a common-mode level at $V_{DD}/2$, which is not sufficient to turn on a transistor in this technology. A body-input gate-clocked comparator has been used as shown in Fig. 6.11. This comparator uses the body terminal of a transistor for signal input and the gate terminal of the same transistor for clock input.

The comparator consists of two stages, the pre-amp and the latch, operating for ϕ LOW and ϕ HIGH respectively. Through the body transconductance of M_1 and M_2 the differential input signal with a CM level of about $V_{DD}/2$ is amplified. The latch stage M_5-M_8 consists of two cross-coupled body-input inverters. The comparator's output is valid for ϕ HIGH.

Non-overlapping clocking between ϕ and $\overline{\phi}$ is not required because the outputs of the latch will not be affected by it inputs once it is latched. However, phase $\overline{\phi}$ should be taken before ϕ in the inverting process so that the load transistors M_3 and M_4 are turned off a bit earlier than M_1 and M_2 to ensure that the pre-amplified signal stored at nodes Q and $\overline{Q}$ gets properly latched.

6.4.3 Operational transconductance amplifiers

The fully differential body-input two-stage OTA from Section 2.1 ([56]) was combined with an automatic biasing circuit to adjust the output CM voltage to about

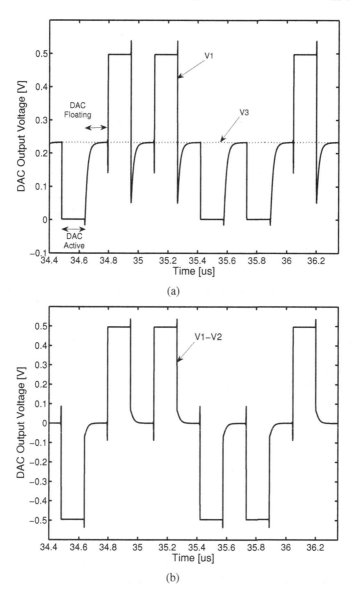

Fig. 6.10: (a) Simulated voltage waveform at the output of the first positive DAC of the modulator (node 1 in Fig. 6.5(a)). (b) The corresponding differential DAC output voltage waveform, i.e, the voltage difference between node 1 and node 2 of Fig. 6.5(a).

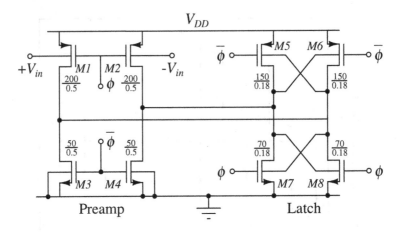

Fig. 6.11: A 0.5 V body-input gate-clocked comparator.

$V_{DD}/2$ against PVT variations for the realization of the OTAs of the modulator. The Miller-compensated two-stage body-input OTA is shown in Fig. 6.12.

The component values shown in Fig. 6.12 are for the first OTA of the modulator. The component values for the other OTAs are the same as the sizes in Table 2.1. The first OTA has been designed for optimal noise performance by increasing the channel length of load transistors ML1 and ML2. A single biasing circuit, redesigned from Section 2.3 ([79]) and presented in Fig. 6.13 showing the component values, is used for all the OTAs.

6.4.4 Clock generation circuit

The clock generation circuit is shown in Fig.6.14. It takes an external low-jitter clock CLK$_{IN}$ as its input and uses body-biased gate-input logic circuits to generate the required four clock phases: phases ϕ and $\overline{\phi}$ for the comparator; and phase ϕ_{DAC} and its slightly advanced version ϕ^a_{DAC} for the RTO DAC. The OR gate ensures that the high interval of ϕ_{DAC} is enclosed by the high interval of ϕ. The $10\%T_s$ delay for ϕ_{DAC} with respect to ϕ is generated by an inverter chain. This compact realization of the 10%Ts delay is sensitive to the supply voltage variation, but system level simulations show that the modulator performance is robust against delay variations up to $\pm5\%T_s$ (see Fig. 6.7). Furthermore, a separate and clean V_{DD} is used for the clock generation circuit to minimize supply noise induced clock jitter.

6.5 Experimental Results

A prototype modulator was designed in a 0.18 μm triple-well mixed-signal CMOS technology using only standard nMOS and pMOS transistors. Metal-insulator-metal (MIM) capacitors and high-resistivity poly resistors were used for their high linearity

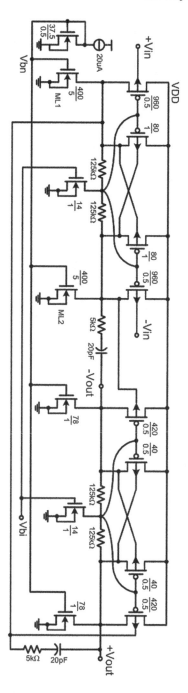

Fig. 6.12: Schematic of the first OTA of the modulator.

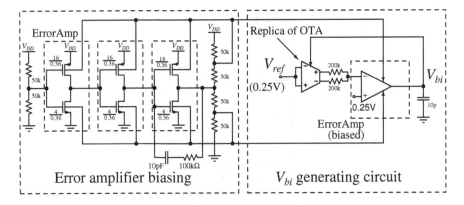

Fig. 6.13: OTA biasing circuits.

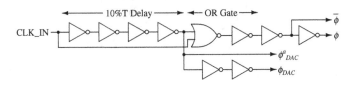

Fig. 6.14: Clock generation circuit.

and small size. Triple-well nMOS transistors were used so that signals or biasing can be applied to the transistor body terminals. Fig. 6.15 shows the die photo. The modulator core occupies an area of 0.6 mm^2.

Six packaged devices available to us were tested and they performed very closely. The typical modulator performance at room temperature is summarized in Table 6.1. For the nominal V_{DD} of 0.5 V a peak SNDR of 74 dB is achieved in a 25 kHz bandwidth from 50 Hz to 25 kHz for a full range differential input signal of 1 $V_{\text{p-p, diff}}$ with a power consumption of 300 μW for the modulator plus 70 μW for the large output buffers to drive the measurement equipment. A modulator output spectrum for an input frequency of 5 kHz, computed off-line from a captured bit stream with Blackman windowing, is shown in Fig. 6.16. The SNDR for varying input levels is presented in Fig. 6.17. Thanks to the RTO technique, the modulator has an excellent distortion performance. The in-band second harmonic is 83 dB below the fundamental (Fig. 6.16), and the third harmonic is 87 dB below the fundamental[3]. The peak

---

[3] The higher second harmonic relative to the third harmonic is possibly caused by mismatches between the positive half-circuit and the negative half-circuit, or by common-mode rejection limitations in the OTAs; we have no experimental data to pinpoint the exact cause of the second harmonic.

Fig. 6.15: Die photograph of the 0.5 V continuous-time $\Sigma\Delta$ modulator.

SNDR was recorded at full input range, implying that the maximum SNDR is likely to be limited by the signal range rather than distortion or noise[4].

Measurements on a test OTA included on the die and which is a replica of the first OTA in the modulator, showed a DC gain of 46 dB, a unity gain bandwidth of 4 MHz with a 25 kΩ load resistor at each output. The measured input referred noise is 33 nV$_{rms}/\sqrt{Hz}$ at 10 kHz and 12 nV$_{rms}/\sqrt{Hz}$ at 1 MHz. The noise performance has been significantly improved compared to the previous design in Section 2.1 ([56]) thanks to using longer channel devices for the load transistors in the body-input OTAs.

Based on the $1/f$ noise data at 10 kHz and the white noise data at 1 MHz, the magnitude of the total input referred noise of the OTA is estimated to be 8.4 μV$_{rms}$ over a bandwidth from 50 Hz to 25 kHz. With this OTA noise magnitude, according to (6.4), the total input referred noise of the modulator is 31 μV$_{rms}$ (single-ended). For a maximum differential input magnitude of 1 V$_{p\text{-}p,\ diff}$, the resulted peak SNR of the modulator would be 78.1 dB.

The measured value of the jitter of the critical clock ϕ_{DAC} is 300 ps peak-to-peak for both rise and fall edges. Assuming the jitter noise is evenly distributed between -150 ps and +150 ps, which is an over-estimation, the jitter limited SNR of a CT $\Sigma\Delta$ modulator with 50% feedback duty cycle is [98]:

---

[4] A suggestion to improve the performance for future designs is to make the peak SNDR occur at input level slightly lower than full scale, if the variation on the resistors and capacitors can be compensated by means like component switching or trimming.

Table 6.1: Modulator performance summary at 25°C

Modulator type	1-bit, 3rd order, continuous-time		
Signal bandwidth	25 kHz (50 Hz - 25 kHz)		
Sampling frequency / OSR	3.2 MHz / 64		
Differential input signal voltage range	1 $V_{\text{p-p, diff}}$		
Supply voltage	0.45 V	0.5 V	0.8 V
SNDR V_{in}=1 $V_{\text{p-p, diff}}$	71 dB	74 dB	74 dB
SNR V_{in}=1 $V_{\text{p-p, diff}}$	76 dB	76 dB	74 dB
Power consumption (total) Sigma Delta Modulator (filter + comparator + DAC) Output buffers	370 μW 300 μW 70 μW		
Active die area	0.6 mm^2		
Technology	0.18 μm CMOS (standard V_T, triple-well nMOS, MIM capacitors, HiRes Poly resistors)		

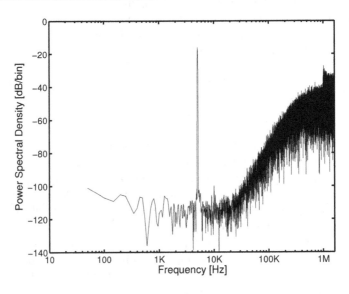

Fig. 6.16: Measured output spectrum at V_{in}=-4dB V_{ref} (V_{ref} = 1 $V_{\text{p-p, diff}}$). Input frequency is 5 kHz, V_{DD} = 0.5 V, and temperature is 25°C. Spectrum computed with 64000-point FFT with Blackman window. Resolution bandwidth is 50 Hz/bin. SNDR = 71 dB.

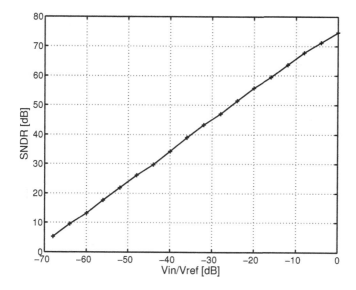

Fig. 6.17: Measured SNDR for varying input level. $V_{DD} = 0.5$ V, temperature=25°C. Signal bandwidth: 50 Hz-25 kHz. Peak SNDR = 74 dB.

$$SNR_{\text{jitter}} = 10 \cdot \log \frac{3}{8E_m^2 \cdot (2\text{BW})^2 \cdot \text{OSR}} \tag{6.6}$$

where E_m is the jitter magnitude (150 ps in this case) for this modulator, and BW is the signal bandwidth. Note that in (6.6) only the jitter from the first feedback DAC is considered. Also, the pulse-delay jitter, which is at least first order noise shaped [91], is ignored. Putting the numerical values to the above equation, the jitter limited SNR of this modulator is 80.2 dB.

The modulator has 78.1 dB and 80.2 dB SNR limits due to thermal-plus-OTA noise and jitter noise respectively. Combining both limits, we obtain an SNR limit of 76 dB, the same as the measured peak SNR.

Figure 6.18(a) shows the measured SNR and SNDR for supply voltage varying from VDD-10% to VDD+60%. The measurements were carried out at room temperature with a fixed input magnitude of 1 $V_{\text{p-p, diff}}$. The modulator's performance is very close to its nominal performance over this wide supply voltage range. Fig. 6.18(b) shows the measured SNR and SNDR for temperature varying from 25°C to 105°C. The measurements were conducted for V_{DD} at 0.5 V and a fixed input level at 1 $V_{\text{p-p, diff}}$. The modulator's performance remains close to, or exceeds, its nominal performance for this temperature range. The modulator's performance at a temperature lower than the room temperature could not be performed due to equipment limitations, but simulations show that the SNDR degradation is less than 1 dB for temperature decreasing from 25°C to -30°C.

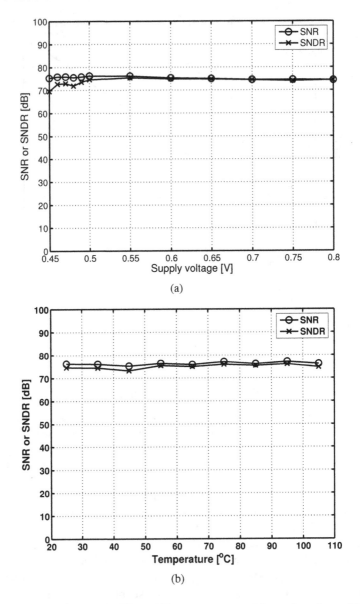

(a)

(b)

Fig. 6.18: SNR and SNDR for $V_{in} = 1$ V$_{\text{p-p, diff}}$ versus (a) supply voltage (T=25°C) and (b) temperature (V_{DD}=0.5 V).

Table 6.2 compares the performance of the proposed modulator with recently reported modulators operating at 1 V or below. The FOM shown in the table is defined as:

$$\text{FOM} = \frac{\text{Power}}{2^{\text{ENOB}} \times 2 \times \text{Bandwidth}} \text{[J/conversion step]} \tag{6.7}$$

where ENOB = (SNDR[dB] − 1.76)/6.02. Compared to other continuous-time modulators operating at 1 V or below, the presented modulator operates at the lowest supply voltage while maintaining an excellent FOM. The last six rows in Table 6.2 include other types of low voltage modulators operating at 0.9 V or below. The presented modulator has the best FOM for modulators operating at 0.8 V or below while not using any special process options.

Table 6.2: Performance comparison

	V_{DD} [V]	$\|V_T\|^\dagger$ [V]	Type	SNDR [dB]	BW [kHz]	Power [μW]	Area [mm^2]	CMOS [μm]	FOM [pJ/conv]
[87]	1	0.3	CT	51	192	1560	2.53	0.5	14.0
[88]	0.9	0.25	CT	50.9	1920	1500	0.12	0.13	1.36
This work	0.5	0.5	CT	74	25	300	0.6	0.18	1.46
[99]	0.5	0.1	SC	39.6	8	75	0.025	0.15	60.1
[42]	0.6	0.33	SRC	77	24	1000	2.88	0.35	3.60
[40]	0.7	0.43/0.38	SO	67	8	80	0.082	0.18	2.73
[85]	0.8	0.55	SC	64	10	60	0.23	0.35	2.32
[86]	0.9	0.5	SO	80	10	200	0.06	0.18	1.22
[37]	0.9	0.62/0.55	SO	62	16	40	0.85	0.5	1.22

SO = Switched Opamp;   CT = Continuous Time;
SC = Switched Capacitor;   SRC = Switched-RC;
$\dagger$: If V_{TN} and $|V_{TP}|$ are different, then the two values are given as $V_{TN}/|V_{TP}|$.

6.6 Conclusions

In this chapter, a design example of 0.5V continuous-time $\Sigma\Delta$ modulator is presented. One of the main challenges in designing a high performance ultra-low voltage modulator is to implement the return-to-zero signaling in the feedback DAC, which traditionally requires switches operating at half the supply voltage (VDD/2). In this chapter, a return-to-open architecture was introduced to realize the return-to-zero signaling without using a switch operating at VDD/2. The method not only enables the ultra-low voltage operation of the return-to-zero feedback DAC, but also suppress the modulator's input referred noise, making it attractive even for higher voltage designs.

Besides the RTO DAC, a body-input gate-clocked comparator was introduced, along with body-input OTAs introduced in Chapter 3, to further enable the ultra-low voltage operation of the CT $\Sigma\Delta$ modulator.

A prototype third order CT modulator employing the proposed techniques has been designed in a 0.18 μm CMOS technology using only standard V_T devices and without any internal voltage boosting. The modulator core consumes 300 μW from a 0.5 V supply, and achieves a peak SNDR of 74 dB over a signal bandwidth of 25 kHz with a sampling frequency of 3.2 MHz. This performance is maintained over a temperature range from 25°C to 105°C and a supply voltage range from 0.45 V to 0.8 V.

The CT $\Sigma\Delta$ modulator showed in this chapter is another example demonstrating that ultra-low voltage operation can be achieved without using special devices but through circuit-level and system-level innovations.

0.5 V Receiver Front-End Circuits[1]

7.1 Introduction

This chapter discusses low-voltage conventional receiver front-end circuits in the Radio-Frequency (RF) range, namely low-noise amplifiers (LNAs) and mixers, expanding on previous work [25, 26, 100–102]. For our study, we chose a future standard CMOS technology nanoscale node, expected by The International Technology Roadmap for Semiconductors [1] to have a supply voltage of 0.5 V, and a transistor threshold voltage of about 200 mV. In the first part of this chapter we consider design of stand-alone communication circuits . Then, we present design and measured results of a 900 MHz integrated receiver front end.

7.2 RF Receiver System-Level Considerations

Figure 7.1 shows a block diagram of a receiver's RF section; each stage is described with its gain (G_i), noise figure (NF_i), and input-referred third-order intercept point $(IIP_{3,i})$, $i = 1, 2, \ldots$.

The overall noise figure and the intercept point of a cascade of receiver stages depend on the gain distribution among them [103]. High gain in the stages closer to the antenna relaxes the noise figure requirements for the stages down the receiver chain, but the linearity suffers. Too much gain in the front can saturate the following stages. The front-end circuits contribute more to the overall noise, while the overall linearity depends more on the back-end circuits.

The range of the possible signal and interferer power levels at the input of the receiver is wide, while maximum voltage swings in the RF and analog circuits decrease with the scaling of the CMOS technology. The high signal-to-noise ratio required for good reception is harder to maintain, and a high level of gain programmability in all receiver circuits becomes more of a necessity.

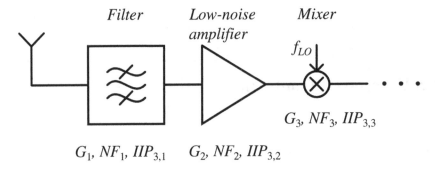

Fig. 7.1: Receiver RF section.

7.3 Low-Noise Amplifiers

7.3.1 Basic Properties and Standard Topologies

A low-noise amplifier (LNA) is typically the first stage with gain in a receiver chain. Its primary task is to amplify the signal with minimal noise contribution, as its noise figure adds directly to the noise figure of the overall system. The LNA must provide enough signal gain in order to reduce the noise contribution of the following stages. The upper limit on the gain is set by the linearity requirement of the overall system.

The LNA serves as a part of the interface to the antenna. It should provide a resistive input impedance, usually 50 Ω, required as a termination for the off-chip filters or baluns, which are placed between the antenna and the amplifier in the receive path.

Another important LNA parameter is reverse isolation. The flow of signal from the output to the input of the amplifier should be small. Good reverse isolation is needed to reduce LO leakage from the mixer to the antenna in homodyne receivers and it improves the LNA's overall stability.

A common-gate LNA topology is shown in Fig. 7.2(a). The resistive part of the impedance for the input matching is provided by the proper choice of the transconductance of the amplifying device. The parasitic capacitance between the input at the source and the output at the drain of the main device is small, providing good reverse isolation. The theoretical noise figure of the common-gate LNA is above 2 dB [103].

Better noise performance is available in the inductively degenerated common-source LNA topology shown in Fig. 7.2(b). The minimum achievable noise figure of this circuit is in fractions of a decibel in nanoscale CMOS technologies. The theoretical limit on the noise figure of the structure is derived in [104]. The input impedance looking into the gate of the transistor with an inductor in its source is partly resistive, which is used for the input matching. A large parasitic capacitance between the gate and the drain of the amplifying device, M_1, results in limited reverse isolation. The isolation can be improved by neutralization or by the addition of a cascode device , M_2, as shown in Fig. 7.2(c).

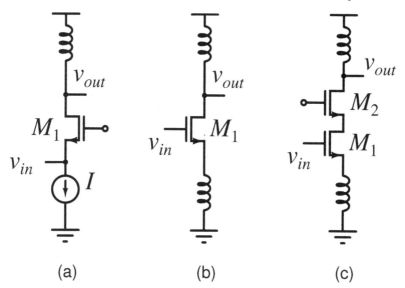

Fig. 7.2: Low-noise amplifiers: (a) common gate, (b) common source with inductive degeneration, (c) cascode common source with inductive degeneration.

7.3.2 Low-Voltage Considerations

To achieve high speed and high performance, the RF circuits require devices biased with a high overdrive voltage $(V_{GS} - V_T)$. Fig. 7.3(a) and Fig. 7.3(b) show the possible bias voltages of the cascode common-source LNA with inductive source degeneration and the common-gate LNA, respectively. The voltage supply of 0.5 V and the transistor threshold voltage of close to 200 mV allow stacking of two devices operating in saturation. The maximum overdrive voltage of the transistors in this mode is about 100 mV, with both devices working in moderate inversion at best.

Transistor body (or back-gate) biasing can be used to reduce the threshold voltages of the transistors, which gives additional headroom. The bodies can be connected to the highest circuit potential, which is the supply voltage. Latch-up is not an issue since 0.5 V is not enough to turn a silicon diode ON [50,61,63]. This technique can be applied when nMOS devices in a well are available.

If higher linearity is required, the number of transistors in the stack can be reduced by folding the cascode structure , as shown in Fig. 7.3(c) [100]. The penalty is a larger circuit area since an additional inductor is required, and a larger power dissipation.

The scaling of the technology leads to an increase in the unity current-gain frequency f_T (i.e., to the faster devices). The circuits may be able to achieve high gain in the RF range without inductors used to tune out the transistor parasitic capacitances [105] (useful for area reduction). However, in an inductorless LNA, an amplifying device and an active load already form a two-device stack. The cascode transistor in the common-source and the current-source transistor in the common-gate con-

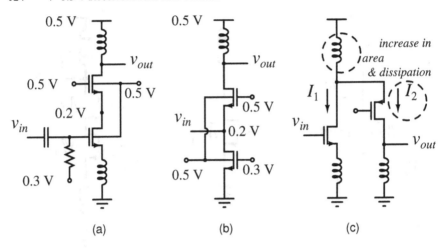

Fig. 7.3: (a) Biasing voltages in a cascode LNA, (b) biasing voltages in a common-gate LNA, (c) stack reduction in folded cascode LNA.

figuration increase the stack height to three. The available overdrive voltages and the output swings are then reduced, resulting in lower linearity. Exclusion of the source-degeneration inductor in the common-source topology requires the use of topologies with higher minimal noise figures. Therefore, in the reduced–voltage-headroom environment, the use of inductors in LNAs becomes increasingly important.

7.4 Downconversion Mixers

7.4.1 Basic Properties and Standard Topologies

A mixer performs frequency translation by multiplication of two signals. A down-conversion mixer comes after the LNA gain in the receive path and usually has more relaxed noise figure (typically above 10 dB [103]), and more stringent linearity requirements.

A double-balanced mixer (based on a four-quadrant Gilbert multiplier [106]) is shown in Fig. 7.4. The RF signal at frequency f_{RF} is applied at the input, while the switches are periodically opened and closed using the local oscillator signal (LO) at frequency f_{LO}. The LO signal is often a rectangular wave. The switching operation is equivalent to the multiplication of the two signals. The desired output of the down-converting mixer is a signal at frequency $f_{IF} = f_{RF} - f_{LO}$, where IF stands for the intermediate frequency. This circuit is widely used in integrated front ends.

Very important in mixers is the port-to-port isolation between all three ports. LO–RF leakage can reach the antenna in the case of poor reverse isolation in the LNA, and can cause DC offsets by self-mixing in homodyne receivers. Homodyne receivers may also suffer from even-order distortion problems in case of RF–IF feed

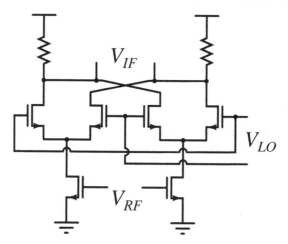

Fig. 7.4: Double-balanced mixer.

through [107, 108]. Finally, a large LO–IF feed through can desensitize the stages following the mixer.

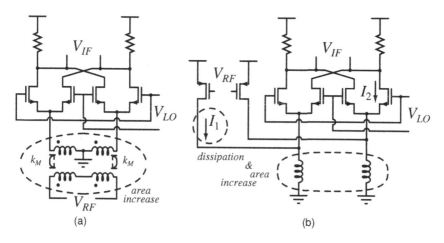

Fig. 7.5: (a) Active mixer with transformer at the RF input, (b) folded Gilbert mixer.

7.4.2 Low-Voltage Considerations

As with LNAs, to maintain high overdrive and thus good performance in the active mixers, the height of the transistor stack within the 0.5 V supply limit has to be kept at two.

The standard Gilbert cell requires three stacked elements with a significant voltage drop on each. The voltage drop across the RF device needs to be high to reduce the effect of its nonlinear output conductance [109, 110]. The LO transistors are most often driven as cascode devices operating in saturation (current switches) for linearity improvement and for higher output resistance. The high voltage drop across the LO devices can be reduced only if they are driven into the linear region. At 0.5 V, this is difficult to do even with a rail-to-rail LO signal since the source potential of the LO transistors is in the mid-rail voltage range. The third voltage drop is needed for a resistive load, or more often for an active load operated in saturation for high output resistance.

Stack reduction can be achieved by coupling the LNA output to the mixer magnetically [101], as depicted in Fig. 7.5(a), or by folding of the Gilbert topology [25], as in Fig. 7.5(b). Additional circuit area for the inductors is required in both methods, and the power dissipation is higher in the folded structure.

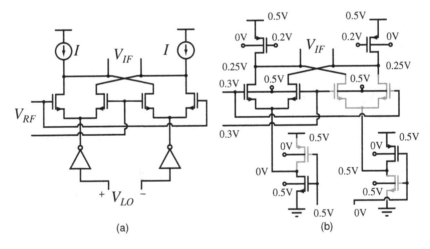

Fig. 7.6: Switched transconductor mixer: (a) simplified schematic (b) biasing voltages for low-voltage operation in one phase of the LO signal.

Suitable for low-voltage operation is the switched transconductor mixer topology, shown in Fig. 7.6(a) [111]. In this circuit, the positions of the switches and the RF-signal amplifying transistors are swapped, compared to the Gilbert cell. The switching is implemented through inverters which can pull the potential at the source of the RF devices to a high or low voltage level. The two RF differential transistor pairs are thus alternately turned ON and OFF periodically. Fig. 7.6(b) shows possible biasing voltages in the switched transconductor mixer for one phase of the square-wave LO signal. The devices that are OFF at that instant are shown in lighter color. When the switching transistors are at the bottom of the stack, they can be hard-switched with the maximum of 0.5 V applied between their gates and sources. The voltage drop across a switch operating in the linear region can be made very small, bringing the source voltage of the RF transistor pair which is ON close to 0 V. The only two devices in the stack are effectively the RF device and the active load. Body biasing can be used to achieve higher overdrive voltages by reducing the transistor threshold voltages, when body contact is accessible.

7.5 900 MHz Receiver Front-End in 0.18 μm CMOS

Figure 7.7 shows the block diagram of a 0.5 V RF front-end operating at 900 MHz, for zero-IF or near-zero-IF receivers [112]. The blocks include an LNA, two quadrature downconversion mixers and their associated LO buffers. The circuits are built in a standard mixed-mode CMOS 0.18 μm process. We used the low-V_T transistors with a threshold voltage of about 200 mV in accordance with the selected technology node (see Section 7.1).

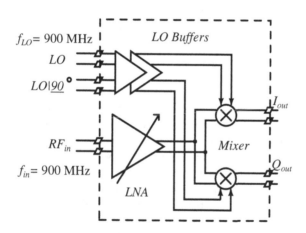

Fig. 7.7: Block diagram of 900 MHz receiver front end.

7.5.1 Design of the LNA

Figure 7.8 shows simplified schematic of an LNA with inductive degeneration , without biasing details. The impedance seen by the source resistance R_s at a frequency ω is:

$$Z_{in}(j\omega) = j\omega(L_g + L_s) + \frac{1}{j\omega C_{gs}} + \frac{g_m L_s}{C_{gs}}, \qquad (7.1)$$

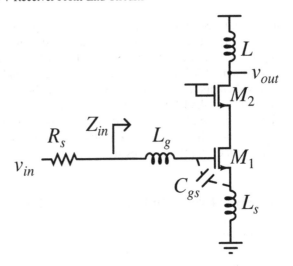

Fig. 7.8: Simplified schematic of the inductively degenerated LNA.

where g_m and C_{gs} are small signal parameters of the input transistor .

The theoretical minimum noise figure of this LNA structure at the frequency ω, for the circuit with a power constraint, is given by [104]:

$$F_{minP} \approx 1 + 2.4\frac{\gamma}{\alpha}(\frac{\omega}{\omega_T}), \qquad (7.2)$$

where γ is the transistor excess noise factor, $\alpha = \frac{g_m}{g_{d0}}$ is the measure of departure from the long-channel regime, and ω_T is the transistor current gain unity gain frequency. The expression takes into account only the noise contributed by the main devices which is most often the dominant noise source in the amplifier. The noise contributions of the cascode device and the loss component contribute less noise in comparison. According to [104], to achieve the above noise figure, the quality factor of the series resonance network (see equation (7.1)) formed at the input to the LNA by the inductors L_g and L_s, the parasitic capacitance of the main devices C_{gs}, and the resistive source and LNA impedances should be $Q_S \approx 4$.

For optimal noise figure, when main device operates in saturation, its width is:

$$W = \frac{3}{2}\frac{1}{\omega L C'_{ox} R_s Q_s}, \qquad (7.3)$$

where C'_{ox} is the oxide capacitance per unit area and L is the length of the device. After this choice of W, the appropriate bias point of the transistors in the amplifier can be chosen for maximum performance. The passive elements values are determined by the selected Q_s.

The voltage gain of the LNA , at the input frequency ω_0 is:

$$A_V = Q_s g_m R_{out}, \qquad (7.4)$$

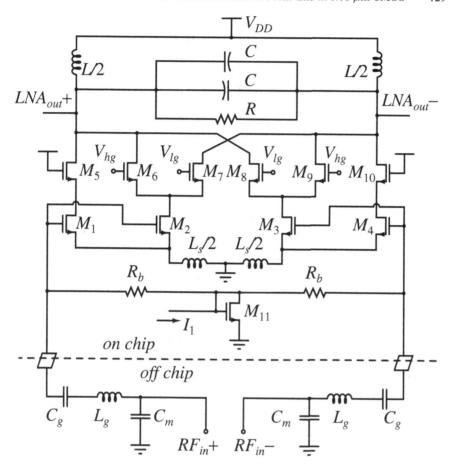

Fig. 7.9: Schematic of the 900 MHz low noise amplifier.

Passive Element	Value	Transistor	Size
L_g	30 nH	M_1, M_4	$\frac{180}{0.24}$
L_s	3.6 nH	M_2, M_3	$\frac{140}{0.24}$
L	9.3 nH	M_5, M_{10}	$\frac{180}{0.24}$
C_m	3.3 pF	M_6, M_7, M_8, M_9	$\frac{140}{0.24}$
C_g	10 pF	M_{11}	$\frac{80}{0.24}$
C	1.5 pF		
R_b	10 kΩ		
R	100 Ω		

Table 7.1: 900 MHz LNA element sizes.

where R_{out} is the output impedance of the LNA at frequency ω_0.

Figure 7.9 shows the schematic of the 900 MHz LNA. The sizes of the elements are listed in Table 7.1. Input transistors M_1-M_4, together with a symmetric differential on-chip inductor L_s and an off-chip matching network, form a matched 100 Ω differential input at 900 MHz. C_m modifies the matching network compared to the circuit in Fig. 7.8 to a quality factor Q_{in} of the input network.

For the main devices M_1-M_4 in our design, we chose the minimum available length for the low-V_T transistors in our technology of L_{min} =0.24 μm, for maximum ω_T. The bias current of the LNA is set through a diode-connected device M_{11}, via control current I_1. The LNA has two gain modes, one of which corresponds to attenuation.

Cascoded devices M_5-M_{10} implement the programmable gain . In the high gain mode, M_6 and M_9 are switched ON, and M_7 and M_8 are kept OFF, by having V_{hg} =0.5 V and V_{lg} =0 V. In the low gain mode, we have V_{hg} =0 V and V_{lg} =0.5 V. M_6 and M_9 are OFF, while M_7 and M_8 are ON, subtracting a fraction of the signal from the signal propagating through M_5 and M_{10}, thus reducing the gain of the amplifier. Transistors M_6-M_9 are of the same size, which helps maintain similar parasitics in the two settings and results in nearly constant input impedance independent of gain mode.

The reason for choosing the above gain setting method can be understood by comparing the two alternatives shown in Fig. 7.10. In both circuits the same ratio between the high and the low gain setting of 8 is chosen. The parasitic capacitance at the intermediate node between the main and the cascode device can lead to signal loss (lower gain). The circuit in Fig. 7.10(b) has the advantage of lower parasitic capacitance C_{par2} compared to Fig. 7.10(a) circuit's parasitic capacitance C_{par1}. The penalty is a higher noise figure for circuit in Fig. 7.10(b), but only in the low gain mode when signal level at the input is high, which is acceptable.

A tuned load at the output of the LNA in Fig. 7.9 is formed using another symmetric differential inductor, L, and the parasitic capacitance of the cascode devices, plus some additional MIM capacitance, C. The load is sized for a simulated peak voltage gain of 16 dB at 900 MHz. 12 dB of the overall LNA gain comes from the passive gain in the input resonance network. For the supply current of 4.5 mA, simulated noise figure and input referred third-order intercept point in the high gain setting were 1 dB and -10 dBm, respectively.

7.5.2 Design of the downconversion mixer

Figure 7.11 shows the schematic of the mixer, AC-coupled to the output of the LNA through capacitors C_{in}. The element sizes are shown in Table 7.2. Input transistors M_{20}-M_{23} form a transconductance stage having as load current sources M_{24} and M_{25}, resistors R_l, and capacitors C_l. The latter elements are used for rough RF filtering at the output of the mixer. Resistors R_o, in combination with current source M_{27}, are used to provide an appropriate voltage drop which biases the drains of M_{24}-M_{25} at a higher potential than their gates, thus reducing the source-drain voltage of those transistors in order to allow operation with a 0.5 V supply [79]. The sources of

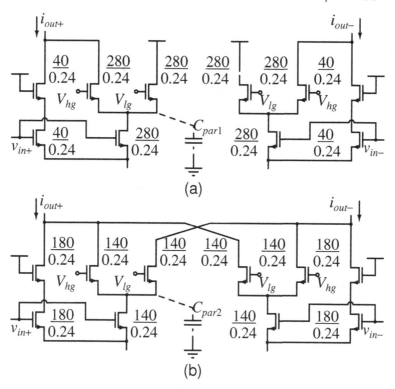

Fig. 7.10: LNA gain settings methods: (a) classical, (b) implemented.

the two input transistor pairs M_{20}, M_{21} and M_{22}, M_{23} are pulled to a high or low voltage level through inverters switched via signals LO_{sq} and $\overline{LO}_{sq}$. Transistors M_{31} and M_{33} are hard-switched, and have a low ON resistance in the triode region, with a voltage drop of less then 25 mV across each of them when conducting current. The DC operating point of the mixer is set via two control currents, I_2 and I_3. Transistor M_{29} in the source of the diode-connected device M_{28} models the voltage drop across the nMOS switches. The gates of the input transistors are fed the bias voltage via resistors R_i.

The action of the two input differential pairs is described in Fig. 7.12. The effective transconductances of the two transistor pairs driving the differential load are denoted g_{m1} and g_{m2}. They are approximated with a trapezoid, taking values g_{m0} and 0 in the two phases of the LO_{sq} signal. The clock signals are assumed to be 50 % duty cycle square waves with period T_{LO} and with finite transition times τ_{sw} between the low and the high level. Using the above approximation, the conversion gain of the mixer can be found to be [111]:

$$CG \approx \frac{2}{\pi} \frac{sin(\pi f_{LO}\tau_{sw})}{\pi f_{LO}\tau_{sw}} g_{m0}(R_l||R_o) = cg_{m0}(R_l||R_o). \qquad (7.5)$$

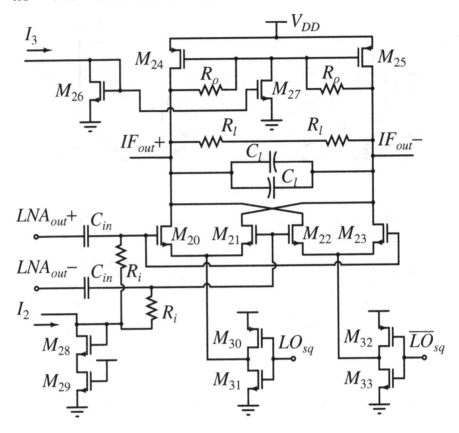

Fig. 7.11: Schematic of the 900 MHz mixer.

Passive Element	Value	Transistor	Size
C_{in}	2.5 pF	M_{20}, M_{21}, M_{22}, M_{23}	$\frac{100}{0.24}$
C_l'	2.5 pF	M_{24}, M_{25}	$\frac{480}{0.24}$
R_i	10 kΩ	M_{26}, M_{27}	$\frac{30}{0.24}$
R_o	2 kΩ	M_{28}	$\frac{25}{0.24}$
R_l	200 Ω	M_{29}	$\frac{90}{0.24}$
		M_{30}, M_{32}	$\frac{288}{0.24}$
		M_{31}, M_{33}	$\frac{144}{0.24}$

Table 7.2: 900 MHz downconversion mixer element sizes.

The thermal noise in the mixer at the differential output is generated by the differential pairs and the output load. The devices in the LO buffers generate thermal noise only in common mode. The resulting noise figure of the mixer is [111, 113]:

$$NF_{SSB} \approx \frac{\alpha}{c^2} + \frac{2(\gamma_{G_m} + r_{g,G_m}g_{m0})g_{m0}\alpha + \frac{2}{R_l} + \frac{2}{R_o} + 2\gamma_{G_{m,out}}g_{m,out}}{c^2 g_{m0}^2 R_s}, \quad (7.6)$$

where γ_{G_m} and r_{g,G_m} are the excess noise factor and gate resistance of the input transistors respectively. $\gamma_{G_{m,out}}$ and $g_{m,out}$ are the excess noise factor and the transconductance of the pMOS devices in the output load. α models the noise folding, and its value for the trapezoid approximation is:

$$\alpha \approx 1 - \frac{4}{3}\tau_{sw}f_{LO}. \quad (7.7)$$

The flicker noise at the differential output in the mixer is generated by the input nMOS transistor pairs and the output pMOS transistor pair, while the devices in the LO buffer produce flicker noise only in common mode.

In simulation, the mixer from Fig. 7.11 achieved 0 dB gain, IIP$_3$ of 3 dBm and noise figure of 20 dB (1/f noise corner was at 200 kHz), while dissipating 1.3 mA from the 0.5 V supply.

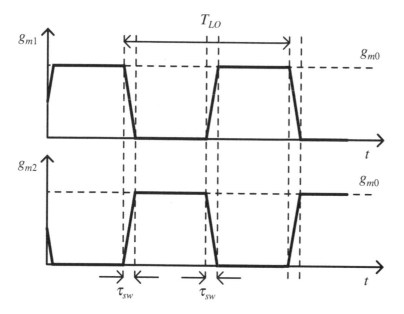

Fig. 7.12: Transconductance of the two input differential pairs of the mixer.

7.5.3 Design of the LO buffers

The sources of the input transistors in the downconversion mixer are driven by complementary square-wave voltages, formed by on-chip buffering of an off-chip generated sinusoidal signal. Such buffering is performed in two stages, of which the second stage is shown in Fig. 7.11, and the first in Fig. 7.13. In each of the complementary buffer sides, DC negative feedback sets the quiescent input-output voltage of the buffers at about half the supply voltage, with all devices active. This is accomplished using the large buffer inverters in cascade with two small inverters, also shown in Fig. 7.13. A differential input signal, LO and $\overline{LO}$, is AC-coupled to the buffers as shown, and causes transitions around the above quiescent point, resulting in a 50 % duty cycle of the square wave signal driving the mixer. The resistors R_{f1} and R_{f2} and the off-chip capacitors C_f decouple DC from AC operation, and the capacitors also serve for frequency compensation of the loop.

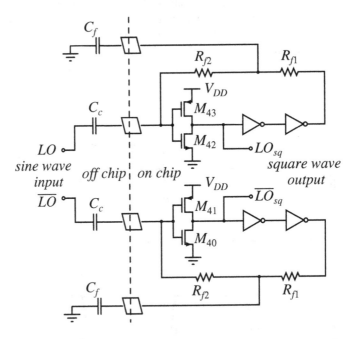

Fig. 7.13: Schematic of the 900 MHz LO buffers.

Passive Element	Value	Transistor	Size
C_c	10 pF	$M_{40}, M_{42},$	$\frac{144}{0.24}$
C'_f	200 pF	M_{41}, M_{43}	$\frac{288}{0.24}$
R_{f1}	50 kΩ		
R_{f2}	50 Ω		

Table 7.3: 900 MHz LO buffer element sizes.

7.5.4 Measurement and results

A chip microphotograph is shown in Fig. 7.14. The chips have been bonded to a test board containing the external input matching components. For characterization, off-chip RF baluns are used to interface to the single-ended measurement equipment. Measurement results are reported from the differential RF input to the differential IF output of the front-end. The highest gain was recorded at the RF input frequency of 875 MHz. Reflection (S_{11}) at the RF input was better than -10 dB in a 60 MHz band around the peak gain frequency point. Fig. 7.15 and Fig. 7.16 show the gain and the reflection, respectively, for both LNA's gain settings.

Figure 7.17 and Fig. 7.18 show measurements used to obtain the input referred 1-dB compression point and the input referred third-order intercept point in the high-gain setting. A summary of the front end's performance at a room temperature is given in Table 7.4. Results are listed for the front end operating at the nominal supply voltage of 0.5 V in both gain modes, and for supply voltage values 10% away from the nominal value with gain set to high. The same values for the circuit's bias currents I_1 and I_2 were used in all cases, with a fixed output common-mode voltage. The latter can be set by a DC feedback loop as in [79], but for the purposes of this testing it was set manually by adjusting bias current I_3. The measured $1/f$ noise corner frequency at the mixer IF output was 200 kHz.

We compare simulation and measurement results for the front-end and its blocks in Table 7.5. The results are obtained for nominal supply voltage and high gain setting at room temperature. The LNA's and mixer's separate performances were extracted (calculated) from the overall performance using noise figure and linearity equations for a cascade of stages [103]. The differential output impedance of the LNA, required for the power gain calculations, was measured (by RF probing) to be completely resistive and close to 100 Ω at frequency of 875 MHz. Table 7.6 compares the results of our work to the other published work.

We have performed measurements over temperature from $-5°C$ to $85°C$, with PTAT currents I_1 and I_2 supplied externally, and output common-mode voltage kept constant as above. The peak variation in performance is recorded at the highest temperature with gain decrease of 2.3 dB, noise figure increase of 1.7 dB and IIP$_3$ increase of 1 dB compared to the results measured at room temperature.

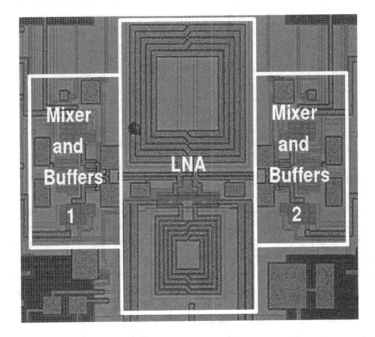

Fig. 7.14: Chip microphotograph of the 900 MHz receiver front end.

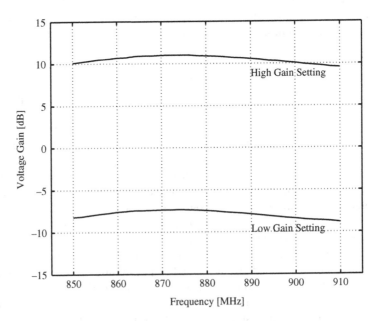

Fig. 7.15: Voltage gain of the front end.

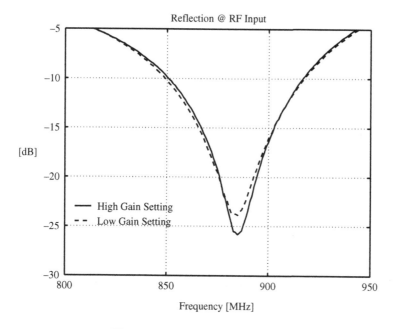

Fig. 7.16: Reflection at the RF input.

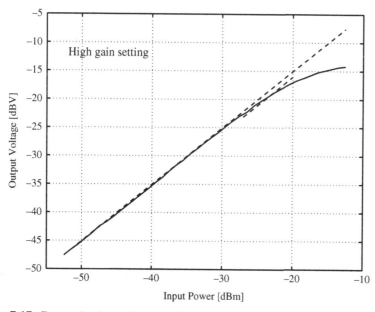

Fig. 7.17: Determination of input referred compression point of the front-end for nominal supply voltage and high-gain setting.

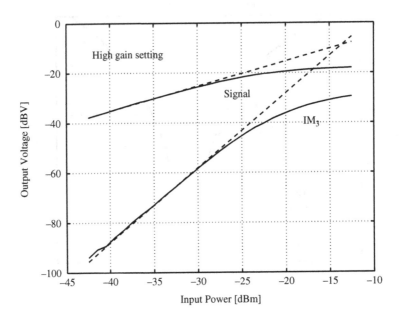

Fig. 7.18: Determination of input referred third-order intercept point of the front-end for nominal supply voltage and high-gain setting.

Supply voltage	0.45V	0.5V		0.55V
Gain setting	High	High	Low	High
Voltage gain @ 875 MHz [dB]	10	12	-6	13
CP_{1-dB} @ input [dBm]	-23	-23	-19	-23
IIP3 [dBm]	-14	-14	-10	-14
DSB NF @ IF=1 MHz [dB]	11	9	26	8
S_{11} (minimum) [dB]	-21	-26	-24	-31
Power dissipation [mW] LNA:	2.2	2.6		3
Mixer (each channel):	0.5	0.6		0.7
Buffers (each channel):	1.4	1.8		2.2
Core circuit dimensions LNA:	650 μm x 350 μm			
Mixer + Buffers (each channel):	400 μm x 250 μm			

Table 7.4: Summary of 900 MHz front-end performance @ 25°.

	LNA		Mixer		Front-end	
	Meas.	Sim.	Meas.	Sim.	Meas.	Sim.
Voltage Gain [dB]	12	16	0	0	12	16
IIP3 [dBm]	-12	-12	3	3	-14	-14
NF [dB]	1.5	1	20	20	9	6

Table 7.5: Comparison of simulated and measured results.

		[25]	[100]	[26]	[101]	[102]	This work [112]
LNA	Supply voltage [V]	0.5	0.6	0.5			0.5
	P_{diss} [mW]	2	2.1	2.5			2.6
	Frequency [GHz]	2	5	2.45/5.25			0.9
	Voltage Gain [dB]	8	11	13.9/8.7			12
	NF [dB]	3.9	3.2	5.1/6.6			1.5
	IIP_3 [dBm]	N/A	-8.6	2.9/3.2			-12
	Technology [nm]	200	90	180			180
Mixer	Supply voltage [V]	0.5			0.6	0.6	0.5
	P_{diss} [mW]	4			1.6	0.8	0.6
	Frequency [GHz]	2			2.5	5.2	0.9
	Voltage Gain [dB]	1.5			5.4	3	12
	NF [dB]	19.5			14.8	N/A	20
	IIP_3 [dBm]	-8			-2.8	-8	3
	Technology [nm]	200			130	180	180

Table 7.6: Comparison with the prior art.

A

Analysis of a Distributed Model for a MOS Capacitor

In this section, we will analyze the distributed model for the MOS capacitor, shown in Fig. A.1. The analysis here is modeled after [71].

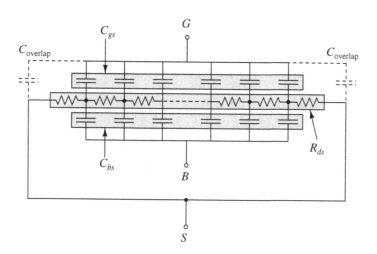

Fig. A.1: Distributed model for a MOS capacitor.

Of particular interest in this analysis, is the gate-source transadmittance, $\overline{Y_{gs}}(jw)$, which is defined as:

$$\overline{Y_{gs}} = \left. \frac{\overline{I_g}}{\overline{V_s}} \right|_{V_g, V_b = 0} \tag{A.1}$$

$\overline{I_g}$, $\overline{V_s}$ are shown in Fig. A.2. To evaluate $\overline{Y_{gs}}$, we can perform a thought experiment as shown in Fig. A.2. $\overline{I} = \overline{I_g} + \overline{I_b}$ is measured with an ammeter, when a signal $\overline{V_s}$ is applied at the drain-source. It can be seen that

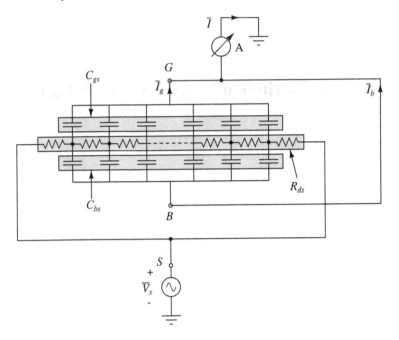

Fig. A.2: Thought experiment to measure $\overline{Y_{gs}}$.

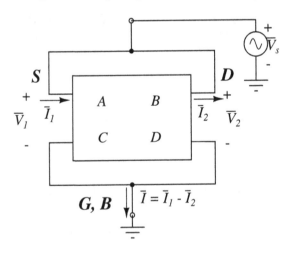

Fig. A.3: Replacing the MOS capacitor with its lumped transmission parameters.

$$\overline{I}_g = \frac{C_{gs}}{C_{gs} + C_{bs}} \overline{I} \tag{A.2}$$

The MOS capacitor in Fig. A.2 can be replaced with a circuit that has the same lumped transmission parameters as shown in Fig. A.3 with the correspondence in the terminals as shown. Then, $\overline{Y_{gs}}$ can be represented in terms of the lumped transmission parameters in the following manner:

From the transmission matrix, we get:

$$\overline{V}_2 = A\overline{V}_1 + B\overline{I}_1 \tag{A.3}$$
$$\overline{I}_2 = C\overline{V}_1 + D\overline{I}_1 \tag{A.4}$$

Applying Kirchoff's equations to the circuit, we get:

$$\overline{V}_s = \overline{V}_1 = \overline{V}_2 \tag{A.5}$$
$$\overline{I} = \overline{I}_1 - \overline{I}_2 \tag{A.6}$$

From (A.3), (A.4) and (A.5), (A.6), we can deduce:

$$\overline{I}/\overline{V}_s = \frac{(1 - A)(1 - D)}{B} - C \tag{A.7}$$

$\overline{Y_{gs}}$ can be obtained from (A.1), (A.2) and (A.7).

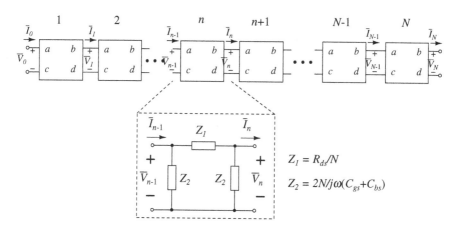

Fig. A.4: Expanding the distributed network into a cascade of N sections, $N \to \infty$.

The lumped transmission network in Fig. A.3 can be broken up into a cascade of N sections, where $N \to \infty$. This is shown in Fig. A.4. In each section, Z_1 and Z_2 are given by:

$$Z_1 = R_{ds}/N \tag{A.8}$$
$$Z_2 = 2N/j\omega(C_{gs} + C_{bs}) \tag{A.9}$$

For the nth section,

$$\begin{bmatrix} \overline{V_{n+1}} \\ \overline{I_{n+1}} \end{bmatrix} = \begin{bmatrix} 1 + \frac{Z_1}{Z_2} & -Z_1 \\ -\frac{Z_1 + 2Z_2}{Z_2^2} & 1 + \frac{Z_1}{Z_2} \end{bmatrix} \begin{bmatrix} \overline{V_n} \\ \overline{I_n} \end{bmatrix} = [M] \begin{bmatrix} \overline{V_n} \\ \overline{I_n} \end{bmatrix}$$

And for the complete network,

$$\begin{bmatrix} \overline{V_N} \\ \overline{I_N} \end{bmatrix} = [M]^N \begin{bmatrix} \overline{V_0} \\ \overline{I_0} \end{bmatrix}$$

Note that $[M]^N$ is the transmission matrix for the entire network. Now let us identify the eigenvalues of the matrix $[M]$. If λ is an eigenvalue, then:

$$\det \|[M] - \lambda[I]\| \equiv 0$$

or,

$$\lambda^2 - 2\lambda(1 + Z_1/Z_2) + 1 = 0 \tag{A.10}$$

(A.10) is the characteristic equation of the matrix $[M]$. If λ_1 and λ_2 are two solutions to this characteristic equation, and hence eigenvalues of $[M]$, then:

$$\lambda_1 \lambda_2 = 1 \tag{A.11}$$
$$\lambda_1 + \lambda_2 = 2(1 + Z_1/Z_2) \tag{A.12}$$

So, to satisfy (A.11),

$$\text{if } \lambda_1 = e^\tau \text{ and } \lambda_2 = e^{-\tau}$$

then to satisfy (A.12),

$$\cosh \tau = (1 + Z_1/Z_2) \tag{A.13}$$

Now we can simplify the definition of $[M]$ using (A.13):

$$[M] = \begin{bmatrix} \cosh \tau & -Z_1 \\ -\frac{\sinh^2 \tau}{Z_1} & \cosh \tau \end{bmatrix} \tag{A.14}$$

From the Cayley-Hamilton theorem,

$$[M]^N = C_0[I] + C_1[M] \tag{A.15}$$

where C_0 and C_1 are constants, and

$$\lambda^N = C_0 + C_1\lambda \tag{A.16}$$

Inserting the specific eigenvalues in (A.16),

$$(e^\tau)^N = C_0 + C_1 e^\tau \tag{A.17}$$
$$(e^{-\tau})^N = C_0 + C_1 e^{-\tau} \tag{A.18}$$

From (A.17) and (A.18),

$$C_1 = \frac{\sinh N\tau}{\sinh \tau} \qquad (A.19)$$

$$C_0 = -\frac{\sinh(N-1)\tau}{\sinh \tau} \qquad (A.20)$$

From (A.15), (A.14) and (A.19), (A.20), we get:

$$[M]^N = \begin{bmatrix} \cosh(N\tau) & -Z_1 \frac{\sinh(N\tau)}{\sinh \tau} \\ -\frac{\sinh(N\tau)\sinh \tau}{Z_1} & \cosh(N\tau) \end{bmatrix} \qquad (A.21)$$

This gives us the transmission matrix for the complete network. In the limit where $N \to \infty$, (A.21) will be the transmission matrix for the distributed-RC network that we started with, as shown in Fig. A.3.

Now from (A.7) and (A.21), we obtain:

$$\overline{I}/\overline{V_s} = -\frac{[1 - \cosh(N\tau)]^2 \sinh \tau}{Z_1 \sinh(N\tau)} + \frac{\sinh \tau \sinh(N\tau)}{Z_1}$$

Simplifying further and using (A.8) and (A.9), we can deduce:

$$\overline{I}/\overline{V_s} = -\frac{2N}{R_{ds}} \cdot \frac{\sinh \tau}{\sinh(N\tau)} \cdot [1 - \cosh(N\tau)] \qquad (A.22)$$

where τ is given by (A.13), (A.8) and (A.9) as:

$$\tau = \cosh^{-1}\left[1 + \frac{j\omega R_{ds}(C_{gs} + C_{bs})}{2N^2}\right] \qquad (A.23)$$

As a sanity check, when $N = 1$, the network reduces to a simple lumped network, with the channel resistance, R_{ds}, shorted out. As a result, we expect the expressions (A.22), (A.13) to give us a purely capacitive result, i.e, $j\omega(C_{gs} + C_{bs})$.

From (A.23), in the limit when $N \to \infty$,

$$\lim_{N \to \infty} \tau = 0$$

From the definition of τ in (A.23),

$$N^2 \sinh^2(\tau/2) = j\omega R_{ds}(C_{gs} + C_{bs})/4$$

In the limit, as $N \to \infty$ and $\tau \to 0$,

$$\lim_{\substack{N \to \infty \\ \tau \to 0}} N^2 \sinh^2(\tau/2) = j\omega R_{ds}(C_{gs} + C_{bs})/4$$

Simplifying further, we obtain:

$$\lim_{\substack{N \to \infty \\ \tau \to 0}} N\tau = \sqrt{j\omega R_{ds}(C_{gs} + C_{bs})} = \alpha \qquad (A.24)$$

From (A.22), in the limit where $N \to \infty$ and also $\tau \to 0$:

$$\lim_{N \to \infty} \overline{I}/\overline{V_s} = \frac{2\alpha}{R_{ds} \sinh \alpha}(\cosh \alpha - 1) = \frac{\alpha^2}{R_{ds}} \cdot \frac{\tanh \alpha/2}{\alpha/2} \qquad \text{(A.25)}$$

This is rewritten, with the help of (A.24) to give:

$$\lim_{N \to \infty} \overline{I}/\overline{V_s} = j\omega(C_{gs} + C_{bs}) \cdot \frac{\tanh \alpha/2}{\alpha/2}$$

where α is given by (A.24).

Using (A.1) and (A.2), we can finally arrive at the desired expression for $\overline{Y_{gs}}$ as:

$$\overline{Y_{gs}} = j\omega C_{gs} \cdot \frac{\tanh \alpha/2}{\alpha/2} \qquad \text{(A.26)}$$

where α is given by (A.24). As a second sanity check, when $\alpha \to 0$, (A.26) gives $j\omega C_{gs}$, which is expected.

For cases where N does not tend to ∞, but is a large integer, note that $N\tau = \alpha$, as defined, as long as $\tau \to 0$. This means that the expression in (A.26) is approximately valid as long as N is large enough such that $\tau \to 0$. Looking at (A.13), this is true as long as:

$$\omega R_{ds}(C_{gs} + C_{bs}) << 2N^2 \qquad \text{(A.27)}$$

(A.27) can be used to determine the minimum number of discrete sections that will be required to perform a reasonably accurate simulation of this distributed network.

References

1. *The International Technology Roadmap for Semiconductors (2004 edition)*, ITRS, 2004, Tech. Rep., http://public.itrs.net.
2. E. Vittoz, "Future of analog in the VLSI environment," in *IEEE International Symposium on Circuits and Systems*, May 1990, pp. 1372–1375.
3. P. R. Kinget, "Device mismatch and tradeoffs in the design of analog circuits," *IEEE Journal of Solid-State Circuits*, vol. 40, no. 6, pp. 1212–1224, Jun. 2005.
4. K. Bult, "Analog design in deep sub-micron CMOS," in *Proc. European Solid State Circuits Conference*, Sep. 2000, pp. 11–17.
5. Q. Huang, "Low voltage and low power aspects of data converter design," in *Proc. European Solid State Circuits Conference*, Sep. 2000, pp. 29–35.
6. B. Hosticka, W. Brockherde, D. Hammerschmidt, and R. Kokozinski, "Low-voltage CMOS analog circuits," *IEEE Transactions on Circuits and Systems I*, vol. 42, pp. 864–872, Nov. 1995.
7. M. Steyaert, V. Peluso, J. Bastos, P. Kinget, and W. Sansen, "Custom analog low power design: The problem of low voltage and mismatch," in *Proc. IEEE Custom Integrated Circuits Conference*, May 1997, pp. 285–292.
8. W. Sansen, M. Steyaert, V. Peluso, and E. Peeters, "Toward sub 1V analog integrated circuits in submicron standard CMOS technologies," in *IEEE International Solid State Circuits Conference, Digest of Technical Papers*, Feb. 1998, pp. 186–187.
9. J. W. Fattaruso, "Low-voltage analog CMOS circuit techniques," in *International Symposium on VLSI Technology, Systems and Applications*, Jun. 1999, pp. 286–289.
10. T. Ohguro, H. Naruse, H. Sugaya, E. Morifuji, S. Nakamura, T. Yoshitomi, T. Morimoto, H. Kimijima, S. Momose, Y. Katsumata, and H. Iwai, "An 0.18 μm CMOS for mixed digital and analog applications with zero-volt-V_{th} epitaxial-channel MOSFETs," *IEEE Transactions on Electron Devices*, vol. 46, no. 7, pp. 1378–1383, Jul. 1999.
11. S. Bazarjani and W. Snelgrove, "Low voltage SC circuit design with low-V_t MOSFETs," in *IEEE International Symposium on Circuits and Systems*, 1995, pp. 1021–1024.
12. E. Sackinger and W. Guggenbuhl, "An analog trimming circuit based on a floating gate device," *IEEE Journal of Solid-State Circuits*, vol. SC-23, no. 6, pp. 1437–1440, Dec. 1988.
13. C.-G. Yu and R. L. Geiger, "Low-voltage circuit techniques using floating-gate transistors," in *Low-Voltage/Low-Power Integrated Circuits and Systems*, E. Sanchez-Sinencio and A. G. Andreou, Eds. IEEE Press, 1999, ch. 5, pp. 133–173.

14. A. Guzinski, M. Bialko, and J. Matheau, "Body driven differential amplifier for application in continuous-time active-C filter," in *Proc. (IEEE) European Conference on Circuit Theory and Design (ECCTD'87)*, 1987, pp. 315–319.

15. F. Dielacher, J. Houptmann, J. Resinger, R. Steiner, and H. Zojer, "A software programmable CMOS telephone circuit," *IEEE Journal of Solid-State Circuits*, vol. 26, no. 7, pp. 1015–1026, Jul. 1991.

16. B. Blalock, P. Allen, and G. Rincon-Mora, "Designing 1-V op-amps using standard digital CMOS technology," *IEEE Transactions on Circuits and Systems II*, vol. 45, pp. 769–780, Jul. 1998.

17. S. Karthikeyan, S. Mortezapour, A. Tammineedi, and E. Lee, "Low-voltage analog circuit design based on biased inverting opamp configuration," *IEEE Transactions on Circuits and Systems II*, vol. 47, no. 3, pp. 176–184, Mar. 2000.

18. J. Huijsing and D. Linebarfer, "Low-voltage operational amplifier with rail-to-rail input and output ranges," *IEEE Journal of Solid-State Circuits*, vol. SC-20, no. 6, Dec. 1985.

19. R. Hogervorst, J. P. Tero, R. G. H. Eschauzier, and J. Huijsing, "A compact power-efficient 3V CMOS rail-to-rail input/output operational amplifier for VLSI cell libraries," *IEEE Journal of Solid-State Circuits*, vol. 29, no. 12, pp. 1505–1513, Dec. 1994.

20. R. G. H. Eschauzier, L. Kerklaan, and J. Huijsing, "A 100-MHz 100-dB operational amplifier with multipath nested Miller compensation structure," *IEEE Journal of Solid-State Circuits*, vol. 27, no. 12, pp. 1709–1717, Dec. 1992.

21. S. Pernici, G. Nicollini, and R. Castello, "A CMOS low-distortion fully differential power amplifier with double nested miller compensation," *IEEE Journal of Solid-State Circuits*, vol. 28, no. 7, pp. 758–763, Jul. 1993.

22. K. Lasanen, E. Raisanen-Ruotsalainen, and J. Kostamovaara, "A 1-V 5 μW CMOS-opamp with bulk-driven input transistors," in *Proc. 43rd IEEE Midwest Symposium on Circuits and Systems*, 2000, pp. 1038–1041.

23. T. Lehmann and M. Cassia, "1-V power supply CMOS cascode amplifier," *IEEE Journal of Solid-State Circuits*, vol. 36, pp. 1082–1086, Jul. 2001.

24. T. Stockstad and H. Yoshizawa, "A 0.9-V 0.5-μA rail-to-rail CMOS operational amplifier," *IEEE Journal of Solid-State Circuits*, vol. 37, no. 3, pp. 286–292, 2002.

25. M. Harada, T. Tsukahara, J. Kodate, A. Yamagashi, and J. Yamada, "2-GHz RF front-end circuits in CMOS/SIMOX operating at an extremely low voltage of 0.5 V," *IEEE Journal of Solid-State Circuits*, vol. 35, no. 12, pp. 2000–2004, Dec. 2000.

26. S. Hyvonen, K. Bhatia, and E. Rosenbaum, "An ESD-protected, 2.45/5.25-GHz dual-band CMOS LNA with series LC loads and 0.5V-supply," in *Radio Frequency Integrated Circuits Symposium, Digest of Papers*, 2005, pp. 43–46.

27. T. Cho and P. Gray, "A 10 b, 20 Msample/s, 35 mW, pipeline A/D converter," *IEEE Journal of Solid-State Circuits*, vol. 30, no. 3, pp. 166–172, Mar. 1995.

28. J.-T. Wu, Y.-H. Chang, and K. L. Chang, "1.2 V CMOS switched-capacitor circuits," in *IEEE International Solid State Circuits Conference, Digest of Technical Papers*, Feb. 1996, pp. 388–389.

29. T. Brooks, D. Robertson, D. Kelly, A. D. Muro, and S. Harston, "A cascaded sigma-delta pipeline A/D converter with 1.25 MHz signal bandwidth and 89 dB SNR," *IEEE Journal of Solid-State Circuits*, vol. 32, no. 12, pp. 1896–1906, Dec. 1997.

30. J. Steensgaard, "Bootstrapped low-voltage analog switches," in *IEEE International Symposium on Circuits and Systems*, vol. 2, May 1999, pp. 29–32.

31. A. Abo and P. Gray, "A 1.5-V, 10 bit, 14.3-MS/s CMOS pipeline analog-to-digital converter," *IEEE Journal of Solid-State Circuits*, vol. 34, no. 5, pp. 599–606, May 1999.

32. M. Dessouky and A. Kaiser, "A 1-V 1-mW digital-audio $\Delta\Sigma$ modulator with 88-dB dynamic range using local switch bootstrapping," in *Proc. IEEE Custom Integrated Circuits Conference*, May 2000, pp. 13–16.

33. U. Moon, G. Temes, E. Bidari, M. Keskin, L. Wu, J. Steensgaard, and F. Maloberti, "Switched-capacitor circuit techniques in submicron low-voltage CMOS," in *6th International Conference on VLSI and CAD*, Oct. 1999, pp. 349–358.

34. J. Crols and M. Steyaert, "Switched-opamp: An approach to realize full CMOS switched-capacitor circuits at very low power supply voltages," *IEEE Journal of Solid-State Circuits*, vol. 29, no. 8, pp. 936–942, Aug. 1994.

35. M. Keskin, U. Moon, and G. Temes, "A 1-V, 10 MHz clock-rate, 13-bit CMOS $\Sigma\Delta$ modulator using unity-gain-reset opamps," in *Proc. European Solid State Circuits Conference*, Sep. 2001, pp. 532–535.

36. A. Baschirotto and R. Castello, "A 1-V 1.8-MHz CMOS switched-opamp SC filter with rail-to-rail output swing," *IEEE Journal of Solid-State Circuits*, vol. 32, no. 12, pp. 1979–1986, Dec. 1997.

37. V. Peluso, P. Vancorenland, A. M. Marques, M. Steyaert, and W. Sansen, "A 900-mV low-power $\Delta\Sigma$ A/D converter with 77-dB dynamic range," *IEEE Journal of Solid-State Circuits*, vol. 33, no. 12, pp. 1887–1897, Dec. 1998.

38. M. Waltari and K. Halonen, "1-V 9-bit pipelined switched-opamp ADC," *IEEE Journal of Solid-State Circuits*, vol. 36, no. 1, pp. 129–134, Jan. 2001.

39. J. Sauerbrey, T. Tille, D. Schmitt-Landsiedel, and R. Thewes, "A 0.7-V MOSFET-only switched-opamp $\Delta\Sigma$ modulator in standard digital CMOS technology," *IEEE Journal of Solid-State Circuits*, vol. 37, no. 12, Dec. 2002.

40. J. Sauerbrey, D. Schmitt-Landsiedel, and R. Thewes, "A 0.5-V 1-μw successive approximation ADC," *IEEE Journal of Solid-State Circuits*, vol. 38, no. 7, pp. 1261–1265, Jul. 2003.

41. D. Chang, G. Ahn, and U. Moon, "Sub-1-V design techniques for high-linearity multistage/pipelined analog-to-digital converters," *IEEE Transactions on Circuits and Systems I*, vol. 52, no. 1, pp. 1–12, Jan. 2005.

42. G. Ahn, D. Chang, M. Brown, N. Ozaki, H. Youra, K. Yamamura, K. Hamashita, K. Takasuka, G. Temes, and U. Moon, "A 0.6V 82 dB $\Delta\Sigma$ audio ADC using switched-RC integrators," in *IEEE International Solid State Circuits Conference, Digest of Technical Papers*, Feb. 2005, pp. 166–167.

43. Y. Tsividis, *Operation and Modeling of the MOS transistor*, 2nd ed.   Oxford University Press, New York, 1999.

44. E. Vittoz and J. Fellrath, "CMOS analog integrated circuits based on weak inversion operation," *IEEE Journal of Solid-State Circuits*, vol. 12, Jun. 1977.

45. P. R. Gray, P. J. Hurst, S. H. Lewis, and R. G. Meyer, *Analysis and Design of Analog Integrated Circuits*, 4th ed.   John Wiley and Sons, New York, 2001.

46. B. Razavi, *Design of Analog CMOS Integrated Circuits*.   McGraw Hill, New York, 1999.

47. D. Johns and K. Martin, *Analog Integrated Circuit Design*.   John Wiley and Sons, New York, 1997.

48. T. Kobayashi and T. Sakurai, "Self-adjusting threshold-voltage scheme (SATS) for low-voltage high-speed operation," in *Proc. IEEE Custom Integrated Circuits Conference*, May 1994, pp. 271–274.

49. V. R. Kaenel, M. D. Pardoen, E. Dijkstra, and E. A. Vittoz, "Automatic adjustment of threshold and supply voltages for minimum power consumption in CMOS digital circuits," in *Proc. (IEEE) Symposium on Low Power Electronics*, 1994, pp. 78–79.

50. M.-J. Chen, J.-S. Ho, T.-H. Huang, C.-H. Yang, Y.-N. Jou, and T. Wu, "Back-gate forward bias method for low-voltage CMOS digital circuits," *IEEE Transactions on Electron Devices*, vol. 43, no. 6, pp. 904–910, 1996.

51. A. L. Coban, P. E. Allen, and X. Shi, "Low-voltage analog IC design in CMOS technology," *IEEE Transactions on Circuits and Systems I*, vol. 42, no. 11, pp. 955–958, Nov. 1995.

52. K. N. Ratnakumar and J. D. Meindl, "Short-channel MOST threshold voltage model," *IEEE Journal of Solid-State Circuits*, vol. SC-17, no. 5, pp. 937–948, Oct. 1982.

53. M.-J. Chen, J.-S. Ho, and T.-H. Huang, "Dependence of current match on back-gate bias in weakly inverted MOS transistors and its modeling," *IEEE Journal of Solid-State Circuits*, vol. 31, no. 2, pp. 259–262, Feb. 1996.

54. S. Narendra, M. Haycock, V. Govindarajulu, V. Erraguntla, H. Wilson, S. Vangal, A. Pangal, E. Seligman, R. Nair, A. Keshavarzi, B. Bloechel, G. Dermer, R. Mooney, N. Borkar, S. Borkar, and V. De, "1.1V 1GHz communications router with on-chip body bias in 150nm CMOS," in *IEEE International Solid State Circuits Conference, Digest of Technical Papers*, 2002, pp. 270–271, 466.

55. K. von Arnim, E. Borinski, P. Seegebrecht, H. Fiedler, R. Brederlow, R. Thewes, J. Berthold, and C. Pacha, "Efficiency of body biasing in 90 nm CMOS for low power digital circuits," *IEEE Journal of Solid-State Circuits*, vol. 40, no. 7, pp. 1549–1556, Jul. 2005.

56. S. Chatterjee, Y. Tsividis, and P. Kinget, "A 0.5-V bulk-input fully differential operational transconductance amplifier," in *Proc. European Solid State Circuits Conference*, Sep. 2004, pp. 147–150.

57. ——, "A 0.5 V filter with PLL-based tuning in 0.18 μm CMOS technology," in *IEEE International Solid State Circuits Conference, Digest of Technical Papers*, 2005, pp. 506–507.

58. K. P. Pun, S. Chatterjee, and P. Kinget, "A 0.5 V 74dB SNDR 25kHz CT $\Sigma\Delta$ modulator with return-to-open DAC," in *IEEE International Solid State Circuits Conference, Digest of Technical Papers*, 2006, pp. 72–73.

59. *Predictive Technology Model*, Nanoscale Integration and Modeling Group, Department of Electrical Engineering, Arizona State University, 2006, http://www.eas.asu.edu/~ptm/.

60. *MOSIS Test Results for IBM 0.13 Micron Runs (8RF-LM, 8RF-DM)*, MOSIS, 2005, http://www.mosis.org/Technical/Testdata/ibm-013-prm.html.

61. J. W. Tschanz, J. T. Kao, S. Narendra, R. Nair, D. Antoniadis, and A. P. Chandrakasan, "Adaptive body bias for reducing impacts of die-to-die and within-die parameter variations on microprocessor frequency and leakage," *IEEE Journal of Solid-State Circuits*, vol. 37, no. 11, pp. 1396–1402, Nov. 2002.

62. J. T. Kao, M. Miyazaki, and A. P. Chandrakasan, "A 175-mV multiply-accumulate unit using an adaptive supply voltage and body bias architecture," *IEEE Journal of Solid-State Circuits*, vol. 37, no. 11, pp. 1545–1554, Nov. 2002.

63. S. Narendra, J. Tschanz, J. Hofsheier, B. Bloechel, S. Vangal, Y. Hoskote, S. Tang, D. Somasekhar, A. Keshavarzi, V. Erraguntla, G. Dermer, N. Borkar, S. Borkar, and V. De, "Ultra-low voltage circuits and processor in 180nm to 90nm technologies with a swapped-body biasing technique," in *IEEE International Solid State Circuits Conference, Digest of Technical Papers*, Feb. 2004, pp. 156–157.

64. S. Chatterjee, T. Musah, Y. Tsividis, and P. Kinget, "Weak inversion MOS varactors for 0.5 V analog integrated filters," in *Symposium on VLSI Circuits, Digest of Technical Papers*, Jun. 2005, pp. 272–275.

65. H. Huang and E. K. F. Lee, "Design of low-voltage CMOS continuous-time filter with on-chip automatic tuning," *IEEE Journal of Solid-State Circuits*, vol. 36, no. 8, pp. 1168–1177, Aug. 2001.

66. A. Baschirotto, F. Rezzi, and R. Castello, "Low-voltage balanced transconductor with high input common-mode rejection," *Electronics Letters*, vol. 30, no. 20, pp. 1669–1671, Sep. 1994.

67. F. Rezzi, A. Baschirotto, and R. Castello, "A 3 V 12-55 MHz BiCMOS pseudo-differential continuous-time filter," *IEEE Transactions on Circuits and Systems I*, vol. 42, no. 11, pp. 896–903, Nov. 1995.

68. A. S. Sedra and K. C. Smith, *Microelectronic Circuits*, 3rd ed. Oxford University Press, New York, 1991, pp. 861–872.

69. G. Ferri and W. Sansen, "A 1.3V op-amp in standard 0.7μm CMOS with constant g_m and rail-to-rail input and output stages," in *IEEE International Solid State Circuits Conference, Digest of Technical Papers*, 1996, pp. 382–383, 478.

70. J. R. Brews, "A charge-sheet model of the MOSFET," *Solid State Electronics*, vol. 21, no. 2, pp. 345–355, 1978.

71. M. S. Ghausi and J. J. Kelly, *Introduction to Distributed-Parameter Networks*. Holt, Rinehart and Winston Inc., 1968.

72. A. J. Scholten, L. F. Tiemeijer, R. van Langevelde, R. J. Havens, A. T. A. Z. Duijnhoven, and V. C. Venezia, "Noise modeling for RF CMOS circuit simulation," *IEEE Transactions on Electron Devices*, vol. 50, no. 3, pp. 618–632, Mar. 2003.

73. *BSIM3v3.2.2 MOSFET Model Users' Manual*, Device Group, Department of Electrical Engineering and Computer Sciences, University of California, Berkeley, 1999, http://www-device.eecs.berkeley.edu/~bsim3/.

74. M. Banu and Y. Tsividis, "An elliptic continuous-time CMOS filter with on-chip automatic tuning," *IEEE Journal of Solid-State Circuits*, vol. SC-20, no. 6, pp. 1114–1121, Dec. 1985.

75. A. Hastings, *The Art of Analog Layout*. Prentice Hall, Upper Saddle River, NJ, 2001.

76. K. Gulati, M. S. Peng, A. Pulincherry, C. E. Munoz, M. Lugin, A. R. Bugeja, J. Li, and A. P. Chandrakasan, "A highly integrated CMOS analog baseband transceiver with 180 MSPS 13-bit pipelined CMOS ADC and dual 12-bit DACs," *IEEE Journal of Solid-State Circuits*, vol. 41, no. 8, pp. 1856–1866, Aug. 2006.

77. M. Ishikawa and T. Tsukahara, "An 8-bit 50-MHz CMOS subranging A/D converter with pipelined wide-band S/H," *IEEE Journal of Solid-State Circuits*, vol. 24, no. 6, pp. 1485–1491, Dec. 1989.

78. M. Nayebi and B. A. Wooley, "A 10-bit video BiCMOS track-and-hold amplifier," *IEEE Journal of Solid-State Circuits*, vol. 24, no. 6, pp. 1507–1516, Dec. 1989.

79. S. Chatterjee, Y. Tsividis, and P. Kinget, "0.5-V analog circuit techniques and their application in OTA and filter design," *IEEE Journal of Solid-State Circuits*, vol. 40, no. 12, pp. 2373–2387, Dec. 2005.

80. J. H. Fischer, "Noise sources and calculation techniques for switched capacitor filters," *IEEE Journal of Solid-State Circuits*, vol. 17, no. 4, pp. 742–752, Aug. 1982.

81. L. Toth, I. Yusim, and K. Suyama, "Noise analysis of ideal switched-capacitor networks," *IEEE Transactions on Circuits and Systems I*, vol. 46, no. 3, pp. 349–363, Mar. 1999.

82. R. Gregorian and G. C. Temes, *Analog MOS Integrated Circuits for Signal Processing*. John Wiley and Sons, New York, 1986, pp. 500–513.

83. R. Schreier, J. Silva, J. Steensgaard, and G. Temes, "Design-oriented estimation of thermal noise in switched-capacitor circuits," *IEEE Transactions on Circuits and Systems I*, vol. 52, no. 11, pp. 2358–2368, Nov. 2005.

84. P. Vorenkamp and J. P. M. Verdaasdonk, "Fully bipolar 120-MSample/s 10-b track-and-hold circuit," *IEEE Journal of Solid-State Circuits*, vol. 27, no. 7, pp. 988–992, Jul. 1992.

85. R. L. Reverend, I. Kale, G. Delight, D. Morling, and S. Morris, "An ultra-low power double-sampled A/D MASH $\Sigma\Delta$ modulator," in *IEEE International Symposium on Circuits and Systems*, 2003, pp. 1001–1004.

86. J. Goes, B. Vaz, R. Monteiro, and N. Paulino, "A 0.9V $\Delta\Sigma$ modulator with 80dB SNDR and 83dB DR using a single-phase technique," in *IEEE International Solid State Circuits Conference, Digest of Technical Papers*, 2006, pp. 74–75.

87. Y. Matsuya and J. Yamada, "1 v power supply, low-power consumption A/D conversion technique with swing-suppression noise shaping," *IEEE Journal of Solid-State Circuits*, vol. 29, no. 12, pp. 1524–1530, Dec. 1994.

88. T. Ueno and T. Itakura, "A 0.9V 1.5mW continuous-time $\Sigma\Delta$ modulator for WCDMA," in *IEEE International Solid State Circuits Conference, Digest of Technical Papers*, 2004, pp. 80–81.

89. E. J. van der Zwan, "A 2.3-mW CMOS $\Sigma\Delta$ modulator for audio applications," in *IEEE International Solid State Circuits Conference, Digest of Technical Papers*, 1997, pp. 220–221.

90. R. H. M. van Veldhoven, B. J. Minnis, H. A. Hegt, and A. H. M. van Roermund, "A 3.3-mW $\Sigma\Delta$ modulator for UMTS in 0.18-μm CMOS with 70-dB dynamic range in 2-MHz bandwidth," *IEEE Journal of Solid-State Circuits*, vol. 37, no. 12, pp. 1645–1652, Dec. 2002.

91. M. Ortmanns, F. Gerfers, and Y. Manoli, "A continuous-time $\Sigma\Delta$ modulator with reduced sensitivity to clock jitter through SCR feedback," *IEEE Transactions on Circuits and Systems I*, vol. 51, no. 5, pp. 875–884, May 2005.

92. R. H. M. van Veldhoven, "A triple-mode continuous-time $\Sigma\Delta$ modulator with switched-capacitor feedback dac for a GSM-EGE/CDMA2000/UMTS receiver," *IEEE Journal of Solid-State Circuits*, vol. 38, no. 12, pp. 2069–2076, Dec. 2003.

93. P. Benabes, M. Keramat, and Kielbasa, "Synthesis and analysis of sigma-delta modulators employing continuous-time filters," *Analog Integrated Circuits and Signal Processing*, vol. 23, no. 2, pp. 141–152, May 2000.

94. J. A. Cherry and W. M. Snelgrove, "Excess loop delay in continuous-time delta-sigma modulators," *IEEE Transactions on Circuits and Systems II*, vol. 46, no. 4, pp. 376–389, April 1999.

95. R. Schreier, "Delta sigma toolbox," http://www.mathworks.com/, MATLAB Cental > File Exchange > Companion Software For Books > Electronics > Delta Sigma Toolbox.

96. J. Yuan and C. Svennson, "High-speed CMOS circuit techniques," *IEEE Journal of Solid-State Circuits*, vol. 24, no. 1, pp. 62–70, Jan. 1989.

97. W. J. Dally and J. W. Poulton, *Digital Systems Engineering*. Cambridge University Press, 1998.

98. V. F. Dias, G. Palmisano, and F. Maloberti, "Noise in mixed continuous-time switched-capacitor sigma-delta modulators," *IEE Proceedings-G*, vol. 139, no. 6, pp. 680–684, Dec. 1992.

99. K. Ishida, K. Kanda, A. Tamtrakarn, H. Kawaguchi, and T. Sakurai, "Managing leakage in charge-based analog circuits with low-V_{TH} transistors by analog T-Switch (AT-Switch) and super cut-off CMOS," in *Symposium on VLSI Circuits, Digest of Technical Papers*, 2005, pp. 122–125.

100. D. Linten, L. Aspemyr, W. Jeamsaksiri, J. Ramos, A. Mercha, S. Jenei, S. Thijs, R. Garcia, H. Jacobsson, P. Wambacq, S. Donnay, and S. Decoutere, "Low Power 5 GHz LNA and VCO in 90 nm RF CMOS," in *Symposium on VLSI Circuits, Digest of Technical Papers*, 2004, pp. 372–375.

101. C. Hermann, M. Tiebou, and H. Klar, "A 0.6-V 1.6-mW Transformer-Based 2.5-GHz Downconversion Mixer with +5.4-dB Gain and −2.8-dBm IIP3 in 0.13-μm CMOS," *IEEE Transactions on Microwave Theory and Techniques*, vol. 53, no. 2, pp. 488–495, Feb. 2005.

102. H. Hsieh, K. Chung, and L. Lu, "Ultra-Low-Voltage Mixer and VCO in 0.18 μm CMOS," in *Proc. IEEE Radio Frequency Integrated Circuits Symposium*, 2005, pp. 167–170.

103. B. Razavi, *RF Microelectronics*. Prentice Hall PTR, Upper Saddle River, NJ, 1998.

104. D. Shaeffer and T. Lee, "A 1.5-V 1.5-GHz CMOS Low Noise Amplifier," *IEEE Journal of Solid-State Circuits*, vol. 32, no. 5, pp. 745–759, Feb. 1997.

105. A. Maxim, R. Poorfard, R. Johnson, P. Crawley, J. Kao, Z. Dong, M. Chennam, T. Nutt, and D. Trager, "A Fully-Integrated 0.13 μm CMOS Low-IF DBS Satellite Tuner," in *Symposium on VLSI Circuits, Digest of Technical Papers*, 2006, pp. 37–38.

106. B. Gilbert, "A Precise Four-Quadrant Multiplier with Subnanosecond Response," *IEEE Journal of Solid-State Circuits*, vol. 3, no. 4, pp. 365–373, Feb. 1968.

107. A. A. Abidi, "Direct Conversion Radio Transceivers for Digital Communications," *IEEE Journal of Solid-State Circuits*, vol. 30, no. 12, pp. 1399–1410, Dec. 1995.

108. B. Razavi, "Design Considerations for Direct-Conversion Receivers," *IEEE Transactions on Circuits and Systems II*, vol. 44, no. 6, pp. 428–435, Jun. 1997.

109. R. van Langevelde and F. M. Klaassen, "Accurate Drain Conductance Modeling for Distortion Analysis in MOSFETs," in *IEEE International Electron Devices Meeting, Technical Digest*, 1997, pp. 313–316.

110. S. Kang, B. Choi, and B. Kim, "Linearity Analysis for CMOS for RF Application," *IEEE Transactions on Microwave Theory and Techniques*, vol. 51, no. 3, pp. 972–977, Mar. 2003.

111. E. Klumperink, S. Louwsma, G. Wienk, and B. Nauta, "A CMOS Switched Transconductor Mixer," *IEEE Journal of Solid-State Circuits*, vol. 39, no. 8, pp. 1231–1240, Aug. 2004.

112. N. Stanić, P. Kinget, and Y. Tsividis, "A 0.5V 900 MHz CMOS Receiver Front End," in *Symposium on VLSI Circuits, Digest of Technical Papers*, 2006, pp. 228–229.

113. M. T. Terrovitis and R. G. Meyer, "Noise in Current-Commutating CMOS Mixers," *IEEE Journal of Solid-State Circuits*, vol. 34, no. 6, pp. 772–783, Jun. 1999.

Index

ANALOG CIRCUITS AND SIGNAL PROCESSING SERIES

Consulting Editor: **Mohammed Ismail.** *Ohio State University*